PATRICK

DOCKYARD ECONOMY
AND
NAVAL POWER

Elibron Classics
www.elibron.com

Elibron Classics series.

© 2005 Adamant Media Corporation.

ISBN 1-4021-7946-4 (paperback)
ISBN 1-4021-1186-X (hardcover)

This Elibron Classics Replica Edition is an unabridged facsimile of the edition published in 1863 by Sampson Low, Son, & Co., London.

Elibron and Elibron Classics are trademarks of Adamant Media Corporation. All rights reserved.

This book is an accurate reproduction of the original. Any marks, names, colophons, imprints, logos or other symbols or identifiers that appear on or in this book, except for those of Adamant Media Corporation and BookSurge, LLC, are used only for historical reference and accuracy and are not meant to designate origin or imply any sponsorship by or license from any third party.

ARMOUR-PLATE ROLLING MILL. MERSEY IRONWORKS. LIVERPOOL.

DOCKYARD ECONOMY

AND

NAVAL POWER.

BY

P. BARRY,

AUTHOR OF "DIRECT WESTERN TRANSATLANTIC TRADE;"
"THE INTERNATIONAL TRADE OF THE UNITED STATES, CANADA, AND ENGLAND;"
"AMERICAN AND INDIAN TRANSIT;"
"SCHEME OF COSTLESS LANCASHIRE EMIGRATION;"
"THE DOCKYARDS AND SHIPYARDS OF THE KINGDOM," ETC.

LONDON:
SAMPSON LOW, SON, AND CO., 47, LUDGATE-HILL.

1863.

[*The Author reserves the right of Translation.*]

LONDON:
PRINTED BY WILLIAM ODHAMS, BURLEIGH-STREET,
STRAND, W.C.

Dedication.

TO

RICHARD COBDEN, ESQ., M.P.;

M. MICHEL CHEVALIER, MEMBER OF THE
INSTITUTE OF FRANCE, ETC.;

AND

M. SUCHÊTE, MEMBER OF THE LEGISLATIVE ASSEMBLY
FOR TOULON;

MY BEST THANKS ARE DUE,

FOR THE KIND FACILITIES AFFORDED ME IN PURSUING MY INQUIRIES INTO
THE DOCKYARD ECONOMY AND NAVAL POWER OF FRANCE.

PREFACE.

Difficulties of Newspaper Writers. Members of the London daily press seldom write books, and for the very good reason that they do not have the time. The week, Sunday included, is too short for them, and although it were enlarged to eight or ten days for their accommodation, they would be no better off, because the same inexorable demands would continue to be made on them. At the close of the last Parliamentary Session, two or three newspaper writers in the prime of life died from overwork. So it is always, and so probably it will be always. The older members of the London press could not, if they tried, reckon up the number of fellow-workers who have fallen at their side, young, and seemingly robust, but physically shattered by the ceaseless strain upon their faculties. Serving the public too well, they killed themselves, and be it said to their honour and the public shame, they were never once remembered. The Editor lives in an atmosphere of care. His assistant, or sub, begins the day at nine o'clock at night, and closes it about the time that early-rising Londoners betake themselves to Covent Garden, the Borough Market, or Billingsgate. Refreshing sleep he enjoys only during his four weeks' annual holiday at Southend, Brighton, or elsewhere. The leader-writer is little better off. At nine or ten o'clock at night the Continental telegrams are received; by eleven o'clock at night he knows what the Commons and the Lords have been saying in Session time; and late on Sunday night important American news is brought by the *Persia* or the *Scotia*. He then has to write his article; and when he has finished at one or two o'clock in the morning, it remains for him to find his way home as best he can. If he does not write on the latest news, he may be required to write against time—that is, to send in his leader by seven, eight, nine, or ten o'clock. Failure is a crime, and the article not being up to the mark is a crime. Last of all, the reporter comes here and goes there, and is a lucky fellow if possessed of sufficient leisure to read the morning papers regularly, or catch a glimpse of "Temple Bar" or "Cornhill." Many members of the London daily press, and the writer among the number, have not yet enjoyed the leisure to read either "Cornhill" or "Temple Bar," although both these popular magazines have now a standing of some years. For such men, therefore, to write books is barely possible. Their beginnings and resumptions are constantly interrupted, and if a dozen or more sides of manuscript happen to be got through at any single sitting, it is always on the understanding that when the matter is in type it will be gone through carefully. Alas! when the proofs are received, the unfortunate newspaper writer happens to have more to do than usual, and all the time and attention that can be bestowed on them is to read them, pencil in hand, in the train or omnibus, on the road to the printer's. No wonder that from such a task most newspaper writers shrink, and that some of the most talented have lived and died without ever once writing anything that bore their name.

<small>My reasons for doing what usually is left undone.</small> My reasons for doing what usually is left undone may be stated in a few words, and it is for the reader to judge whether they justify the hastily written pages that are before him. First, I think the time has come when there should be plain speaking about naval matters; and second, that not a few of those interested in naval matters will like to hear a newspaper writer making a clean breast of it. Such are my reasons. During the past three years I have filled, and still fill, the position of naval writer on the staff of a morning paper, writing on the manning of the Navy, ordnance, fortifications, the construction and equipment of ships of war, the claims of the officers and seamen of the mercantile marine, the position of the officers of the Royal Navy, the Dockyards, the Admiralty, &c.; and to that position I was recommended by the highest authority on shipping

matters in this country.* In the course of these three years more especially, I have contracted many friendships among dockyard officials and naval officers of great scientific attainment as well as of high rank, and those friendships are far from being exclusively English. It has been, and continues, my good fortune to know American, French, and Russian officers and gentlemen to whom naval matters are a constant study. Then the great private establishments of the country have been always open to me, and the privilege freely made use of has necessarily brought me into contact with practical men of all classes. With the humblest as well as with the highest I have long enjoyed the advantage of comparing notes; and if Mr. Fairbairn will recall the circumstance to his recollection, the talking individual who lay beside him in the bottom of a ditch at Shoeburyness while the 300-pounder was fired at the rigid target, was the writer. Familiar, therefore, with my subject in all its aspects and bearings, and accustomed to think closely and carefully of it, I am much mistaken if opinions so formed will

---

* Mr. Mitchell will, I am sure, excuse the publication of the following letter, because it will be seen presently that one of the difficulties inseparable from attacking our corrupt navy system is the inability of doing so with authority:—

"Shipping and Mercantile Gazette,
54, Gracechurch-street, London,
4th January, 1860.

"MY DEAR ———

"Mr. Barry, the bearer of this note, is the gentleman I recommended to you as likely to serve you in the matter of maritime articles. He has written a great deal on the various shipping questions, and I think his productions will give you satisfaction.

"Believe me,
"Very faithfully yours,
"WILLIAM MITCHELL."

not be welcomed at this juncture, although not connected and expressed in the polished periods which, under other circumstances, their importance justly claims for them. Of right, I think a hard-working man of letters, whose faculties are his sole estate, and whose daily bread does not come to him, and those depending on him, by writing books, may write books on his own terms, just as a hard-working member of Parliament may or may not entertain his constituents with speeches in the recess, or in the free and easy manner in which a Cabinet Minister expresses himself after whitebait at Greenwich.

<small>The origin of the Book and its further defence.*</small> The book in the reader's hands is a mere sequel to a pamphlet published by me some months ago,† in which was embodied the substance of my reports on the condition

---

* I publish the following letter, in order that the non-appearance of the fine photograph frontispiece of the *Minotaur* frigate, 11½ by 4½, and the five other views of the Thames Shipbuilding Company's Works, may be explained:—

"The Thames Ironworks and Shipbuilding Company (Limited), Orchard-yard, Blackwall, E.,

October 9th, 1863.

"Dear Sir,

"We have read over the first 100 pages of your proposed publication, and we regret to find that there is so much personal abuse of the dockyard officers, which we consider entirely undeserved, and charges brought against individuals which we believe to be entirely false; in fact, the style of the whole is so scurrilous, that we must decline having anything to do with the work, and request that under these circumstances no reference whatever may be made to us in it.

"When you first named to us your intention to publish such a work, we especially stipulated that, while attacking the system which we regard as open to many grave objections, there should be no personal abuse of a body of men whom we know to be both thoroughly skilled in their profession, and gentlemen besides; and we are very much disappointed to find that you have so entirely disregarded this understanding.

"I remain, dear Sir,

"Yours truly,

"John Ford, for Self and Company, Directors."

† *The Dockyards and Shipyards of the Kingdom.*

of the Dockyards. I had witnessed the deplorable condition of the Dockyards and the Navy, and Admiral Robinson having, in an unguarded moment, most wrongfully attacked the Shipyards and the principle of private enterprise, I, fresh from the Dockyards, came to their aid. So little, however, was it then possible for me to do in the total absence of leisure, that I mentally resolved on following the matter up. And, necessarily, an undertaking so nearly affecting the interests of one of the great branches of national industry led not only to frequent conversations with those among whom I am always moving, but to inquiries as to the extent to which the effort was likely to receive support.* This is the truth, the whole truth,—there being nothing to conceal; for to the independent spirit in which the book is written, and the fairness that is maintained as between the great private interests, is the hostility of the Thames Shipbuilding Company wholly owing. Until Captain Ford saw the proof of what I had said about the Millwall Company all was well, and it was a pleasure to him, in occasionally sipping wine after luncheon, to recount to me his doings in Turkey when he and Sir Baldwin Walker were in the service together, and his Majesty the Sultan stood something like £500 for a first-rate joke. But the moment it was seen that the Millwall Company were to have justice done them† an unexpected storm burst. The Thames

---

* The cost of the photographs for the edition, inclusive of those of the Thames Works, is a matter of no less than £400.

† It is due to myself and the Millwall Company to state that the Millwall Company are perfectly indifferent as to what is said about them or other people, and that they distinctly told me so.

Company would not take a single copy of the book if Harrison was to be praised in that way, or if rolled armour-plates were to be pronounced superior to hammered plates; and so on.* This was on the 7th October, four days after the first six sheets of the book were handed to Captain Ford at luncheon, in the presence of the captain of the Spanish ironclad now building in the Thames Company's yard,—we all at the time enjoying ourselves. The captain of the Spanish frigate will very well remember Captain Ford's question if the six sheets (ninety-six pages) then handed to him were the whole book, and my statement that as many as two hundred pages had been worked off and would be sent to him as soon as the drying and pressing processes had been undergone at the binder's.† From the 7th to the 9th October I did not meet Captain Ford, and on the evening of the last-named day I received the extraordinary letter that has just been read. But turning from this unpleasant topic,—neither in the first 100 pages nor in the second 100 pages will the reader find any personal abuse of the dockyard officers; indeed, on more than one occasion it is expressly stated that the system, and not the officers, is to be understood as to blame. The charge of scurrility is met as easily. No man, I affirm, can know the dockyard

---

* In the vain hope of conciliating Captain Ford I actually, among other passages, struck out a deserved compliment to Mr. Charles Henwood, the Millwall Company's Naval Architect, that gentleman having sent in a design for the *Warrior*, of which the Admiralty had spoken highly. I have requested the printer, Mr. Odhams, to preserve the ineffectually altered proof.

† The Admiralty are to understand that the substance of these pages was at the time known to Captain Ford; a fact, indeed, that is confessed in Captain Ford's letter.

and naval system of this country without being at a loss for words sufficiently scurrilous to convey a proper sense of his indignation. It is a system of fraud and robbery; and I have yet to learn the necessity of speaking of these in dignified or polished terms. Let Parliament appoint a dockyard and Admiralty corruption committee, and there is abundant reason to believe that, whether such committee is appointed by the Lords or Commons, the crimes of the Dockyards will be censured in even stronger terms than any that I have used. I could myself speak of the construction of one ironclad frigate by reason of the *friendship* of one high retired official.

<small>The subject-matter of the Book.</small> The first chapter acquaints the reader with the appearance and condition of the Dockyards, discusses the question of defended and undefended dockyards, traces the growth of the dockyard towns, discloses the administrative iniquities that prevail, and institutes comparisons between the English, French, and American dockyards. The second chapter deals with the question of the organisation of masses of labourers, points out the conditions on which success depends, and shows that one and all of the conditions are absent from the dockyard system. In that chapter are also traced the fluctuations of dockyard labour, and the progress that is making towards the general adoption of labouring contracts, which originally were peculiar to America, but since the adoption of iron shipbuilding have been transplanted here.

PREFACE. xiii

The third chapter is devoted to the question and details of dockyard manufactures. The fourth chapter takes up the hitherto forbidding problem of shipbuilding, and conclusively establishes the fact that the subject is overlaid with an all but incredible amount of error. Sufficient prominence is given to the wrong-doing both of the theorists and the practicals, and to the imperfections of the practice of the dockyards. The fifth chapter lays down the only sound principles of naval power, and combats the well-known popular fallacies. The sixth chapter is devoted to the position of the Powers, and shows that with moderate forethought England has nothing to fear from France, but possesses a dangerous rival in the United States. The remaining chapters give prominence to a few of the great iron and iron shipbuilding firms, so that the most timid may be assured that the maintenance of the navy may safely be committed to private hands. For one ship of war that the dockyards could build or repair, the private firms could build or repair a hundred, if not more.

The conclusions of the Book.
The conclusions of the book admit almost of expression in a single sentence. Substantially they are nothing more than this: Remodel the whole Dockyard system, and maintain the Navy on an effective war footing by at all times rendering the private enterprise of the country available for its support. At present the great resources of the country are practically unavailable for naval war, and the danger is that before they could become available

during war, the *prestige* and honour of our flag would temporarily be destroyed. For naval war the country was never more unprepared than it is at present, and experienced naval officers tremble at the prospect. Some of them go so far as to assert that we are fifty years behind the United States, but this is exaggeration. That, however, we are shamefully weak admits of no dispute, and that we cannot ever possibly be strong under the existing system of naval administration, is as true as any axiom in mathematics. Since the time of Henry VIII. the axe has not been once fairly laid to the dockyard tree, and trunk and branches are knotted, blasted, and unsightly. On the lifeless dockyards the treasure of this great country is poured out in vain. And yet we are a commercial people. Half a dozen of our business men will successfully manage dealings of an aggregate of as many millions as are voted annually for the Navy by the House of Commons. Half a dozen of our great capitalists will even take in hand the property of an amalgamated line of railway representing a value equal to the receipts of Mr. Gladstone, and make it answer every intended purpose. But still such men and the public tolerate the grossest jobbery and incompetency every year in Whitehall. As far as the Navy is concerned, we might as well be sunk in the lowest depths of ignorance, and practically there might as well be no Parliament; for the £250,000 a week, the £1,000,000 a month, and the £12,000,000 a year are disbursed with neither decency nor guarantee. This book seeks the complete overthrow of the whole system and its reorganisation on sound fighting and business principles.

It is scarcely necessary to add that those whose private establishments are named are in no way responsible for the book nor for the accuracy of the published statements; and that were I by name to thank naval officers, dockyard officers, and the naval architects and others of the private yards to whom I am indebted for my more important facts, I should be doing them a grievous injury. If it is thought that I have given undue prominence to what the reader may regard as a squabble between Captain Ford and myself, my answer is that I have been left without choice, Captain Ford I believe having, with the first ninety-six pages in hand, made an active canvass of the shipyards, &c., *against* the book.

London, 1st December, 1863.

CONTENTS.

Chapter I.

THE DOCKYARDS.

Their present Appearance, 1. Their present State, 3. The Building Slips, 4. The Storekeepers' Sheds, 6. The Workshops, 8. The first Intention of the Dockyards, 9. Consequence of this first Intention, 11. One Source of Mystification, 12. Confirmation of the Value of the Undefended Dockyards, 13. The Defended Dockyard Theory, 14. Dangerous tendency of the Theory, 15. The cognate Trades, 16. The growth of the Dockyard Towns, 17. The position of the Dockyard Towns, 18. No hardship in this Judgment, 19. The Dockyards not wholly unproductive, 20. Dockyard Administration, 23. The Testimony before the last Commissioners, 24. Administrative Cobbling, 26. The Board Inspections, 28. Co-operation Morning Meetings, 30. Piety and Uncharitableness, 32. Pious Master Shipwrights, 35. The French Dockyards, 36. Our System in comparison, 37. French and English Workmen, 39. The American Dockyards, 40. What the American Secretary does not do, 42. American and English Workmen, 43. NOTES: Summary of Dockyard Work in progress, 4. Storekeepers, 6. Workshops, 8. Recommendation against the Right to Vote, 20. Extraordinary state of the Accounts, 23. Morning Meetings, 30. Sunday Labour, 32. Duties of the American Secretary, 41.

Chapter II.

DOCKYARD LABOUR.

The Difficulty of Organising Labour, 45. The Experience of Contractors, 46. The Dockyard Labour Problem, 48. Objections met: the Case of Russia, 50. Admiralty Conciliation, 51. Admiralty Force and Intimidation, 54. Admiralty Supervision and Common Sense, 55. The existing Labour System, 56. Task and Job Work in the Dockyards, 58. Day Work and Day Pay in the Dockyards, 59. What these Failures prove, 60. Limited Earnings, 62. Additional Abuses, 64. Superannuation, 66. Originally a Bribe, 68. Its gross Injustice, 70. Admiralty Disposition to effect a Change, 71. English Shipyard Labour, 72. The Proximate Cause of Change, 73. The Labour Revolution, 76. The American Shipyards, 78. The New York Experience of an English Dockyard Officer, 80. Work on the American Lakes and Mississippi, 82. The Harmony of Contract Interests, 83. Applicability of the System to the Dockyards, 85. NOTES: Task and Job Report, 58. Day Work and Day Pay Report, 59. Limited Earnings Report, 62. Superannuation Report, 66. Wood and Iron, 73.

Chapter III.

DOCKYARD MANUFACTURES.

Sir George Lewis's Principle of Manufactures, 87. The application to the Army, 88. Enfield, 89. Elswick, 94. The Dockyards governed by no Rule, 95. One of the Dockyard Objections to Contract Supplies, 96. One of the Dockyard Objections to Contract Shipbuilding, 97. The Mast-houses, 98. The Boat-houses, 100. The Capstan-houses, 101. The Joiners' Shops, 101. The Plumbers' Shops, 102. The Wheelwrights' Shops, 103. The Millwrights' Shops, 104. The

CONTENTS.

Roperies, 105. The Sail-lofts, 107. The Colour-lofts, 108. The Rigging-houses, 108. The Lead-mill, 109. The Paint-mill, 109. The Metal-mills and Foundry, 109. The Blockmakers' Shops, 110. The Trenail-houses, 110. The Oarmakers' Shops, 111. Caulkers' and Pitch-heaters' Shops, 111. The Turners' Shops, 112. Locksmiths' Shops, 112. The Foundries, 112. Hosemakers' Shops, 113. The Painters' Shops, 113. The Condensors' Shop, 114. The Pump-house, 114. The Fire-engine Shops, 114. The Smitheries, 115. The Steam-hammer Shops, 115. Conclusions, 116. NOTES: Admiral Robinson's Charges, 91. The Contract Gunboat Report, 98. Joiners' Shops Report, 101. Wheelwrights' Shops Report, 103. Millwrights' Shops Report, 104. Roperies Report, 105. Estimate of Excess Stores, 107. Lead-mill Report, 109. Metal-mills Report, 109. Blockmakers' Shop Report, 110. Oarmakers' Shop Report, 111. Painters' Shops Report, 113. Smitheries Report, 115.

CHAPTER IV.

DOCKYARD SHIPBUILDING.

Scientific Uncertainty, 117. The Problem of Shipbuilding, 119. The Method of Solution followed, 120. What might be done, 121. The Dockyards in the main to blame, 123. Effects of such a State of Things, 126. Shipbuilding under the Practical *Régime*, 128. Discovery of the 1859 Committee, 130. Shipbuilding under the Theorists, 131. Serious Defects in the Contract Ironclads, 135. The want of Small Ships, 136. The Case of the *Dalhousie*, 137. The System Practically, 138. The Power vested in the Controller, 140. The Peculiar Duty of the Constructor, 142. The Little that the Superintendent has to do, 142. The Red Tape Routine before anything is done, 143. The unwieldy Gangs, 145. Shoaling: the Farce, 145. The Usage of Precedents, 146. The Unpractical Officers, 147. The Jack-of-all-trades Character of the Shipwrights, 148. The Admission of Inefficiency, 149. Shipyard Shipbuilding, 150. Designing Merchant Ships, 150. Contracts with Workmen, 151. Antagonism between Employer and Employed, 152. NOTES: Extracts

CONTENTS.

from Sir William Symonds' Memoirs, 117. The *Enterprise* and *Naughty Child*, 119. Extracts from Sir William Symonds' Memoirs, 122. Discrepancy in the Cost of Dockyard Ships, 130. Charges of a Mechanic against the *Achilles*, 131. Contrary View: the *Delhi*, 132. *Times* Statement about the Dock of the *Achilles*, 134. Pliability of Dockyard Surgeons, 145.

CHAPTER V.

NAVAL POWER.

The generally assumed Elements, 154. The Superior Claims of Ships, 156. The surpassing Claims of Dockyards, 157. The obvious Inadequacy in Dockyards, 158. Comparison with France in Dockyards, 159. The obvious Muddle about Ships, 160. Comparison with France in Ships, 162. The obvious Error about Seamen, 163. Comparison with France in Seamen, 165. The real Elements of Naval Power: Readiness, 165. Effective Readiness, 167. Organised Readiness, 169. The second real Element of Naval Power: Resource, 172. Resource in Stores, Docks, and Seamen, 174. Immaterial whence Resource is derived, 177. The third real Element of Naval Power: Endurance, 178. Colonies neither a Strength nor Weakness, 179. Probable Immunity of Colonies in Modern War, 181. Conditions of Privateering, 183. Prizes and their Crews, 184. NOTES: Disabilities of Navy Seamen, 163. Sir John Hay's Promotion and Retirement Scheme, 171.

CHAPTER VI.

POSITION OF THE POWERS.

England's Dockyards, 186. The inexpensive means of placing the Dockyards on a War footing, 187. What French *forçats* can do English convicts may do, 189. English Ordinaries, 191. What the Ordinaries are, 192. Where established, 192. Admiralty

CONTENTS. xxi

Regulations respecting them, 193. Ordinaries partake of the Character of Duplication, 194. Objections to the System: Waste, 196. Our Neighbour's System, 197. England's Shipyards and Workshops, 198. Fallacy to be guarded against, 199. French Dockyards, 200. French Strength and Weakness, 202. Algeria, 205. The French Shipyards, 206. America the Dangerous Rival of England, 207.

CHAPTER VII.

THE THAMES IRONWORKS AND SHIPBUILDING COMPANY (LIMITED), BLACKWALL.

Annual Capability, 209. Locality and Plan of the Works: Middlesex, 209. The Essex side, 210. Railway and other Facilities of the Works, 213. No such Facilities possessed by the Dockyards, 213. The Capabilities in excess of those of all the Dockyards, 215. Peculiarity of the Work, 216. The Works chiefly designed for Shipbuilding, 217. The Working System, 218. The Machinery of the Works, 220.

CHAPTER VIII.

THE MILLWALL IRONWORKS AND SHIPBUILDING COMPANY (LIMITED).

Annual Capability, 221. The Direction of the Company, 221. Employer and Employed, 223. Peculiarity of the Works, 224. The Model Room, 225. The Building Yard, 225. The Forge, 227. Rolling-mills and Rolling, 228. The Armour-plate and Battery Mills, 229. Imperfection of this Outline, 230.

CHAPTER IX.

THE MERSEY STEEL AND IRON WORKS, LIVERPOOL

Position and Extent of the Works, 232. Puddle Steel and Furnaces, 234. Furnaces and Machinery, 234. The Forge

Department, 235. Wrought-iron Cannon Manufacture, 236. The Forge Appliances, 238. Stupendous Hammer, 238. The Engineering Shop, 240. The Rolling and Armour-plate Mills, 242. The Heating Furnace, 243. Employer and Employed, 245.

Chapter X.

THE ATLAS WORKS, SHEFFIELD.

Origin and Progress of the Works, 247. The Old Planing Shop, &c., 249. The New Rolling-mill, 252. The Planing and Slotting Shop, &c., 253. What the Admiralty overlook, 254.

Chapter XI.

THE PARKGATE IRONWORKS, SHEFFIELD.

Origin and Progress of the Works, 256. Extent and Facilities, 257. The Lesson taught by the Works, 259.

Chapter XII.

THE ENGINEERING ESTABLISHMENTS.

Maudslay, Sons, and Field, 261. Messrs. John Penn and Son, 267. The Works, 269. The Albion Ironworks, 271. Rennie's History, and Progress of the Firm, 272. The Floating Docks, 275. The Phœnix Foundry, Liverpool, 276. Progress of the Establishment: Guns, 277. Land and Marine Engines, &c., 278. Contract Steamers, 279. The Blackwall Ironworks: Stewart, 280. Engine and Boiler Works, 281.

Chapter XIII.

SHIPBUILDING INNOVATIONS.

The National Company for Boatbuilding by Machinery (Limited), 283. Machine Boatbuilding, 285. Advantages of Machine-made Boats, 286. Utility in the Navy, 287. Success predicted, 288. Machinery Employed, 289. Tripod Masts, 290. Submarine Batteries, 291. The *Connector* Experimental Ship, 292.

Chapter XIV.

THE THAMES SHIPPING INTERESTS.

Deptford-green Dockyard, 294. The Capabilities of Mr. Lungley's Yard, 296. Messrs. Samuda's Yard, Millwall, 297. Messrs. James Ash and Co.'s Yard, Cubitt Town, 299. NOTES: Mode of Interior Fitting known as Unsinkable Shipbuilding, 294. "The Dockyards and Shipyards of the Kingdom," 297. Vessels Built from the Designs of Mr. James Ash, 299.

Chapter XV.

THE MERSEY SHIPPING INTERESTS.

The Birkenhead Ironworks, 301. The Lesson of the Messrs. Laird's Works, 302. Messrs. Jones, Quiggin, and Co., Shipbuilding Yard, Liverpool, 302. Messrs. Thomas Vernon and Son's Iron Shipbuilding Yard, Liverpool, 304. W. C. Miller's Shipbuilding Yard, Toxteth Dock, Liverpool, 306. The Britannia Engine Works, Birkenhead, 307. Woodside Graving Dock Company (Limited), Birkenhead, 308. G. R. Clover and Co., Liverpool, and Private Graving Docks and Building Yard, Birkenhead, 310, Messrs. W. H. Potter and Co.'s Shipbuilding Yard, Baffin-street, Queen's Dock, and Blackstone-street, near Sandon Graving Docks, Liverpool, 311. Cato, Miller, and Co., Brunswick Forge and Ironworks, 312.

PHOTOGRAPHIC PREFACE.

Photography is to blame for the bad views, and to be praised for the good views that appear in this volume. The perfection and the imperfection of the art are given side by side. True, the fogs of November obscure or deny the light that is the chief condition of the artist's success, but even in the worst of all the English months there are occasional blinks of sunshine, and something like sameness of result might for that reason have been looked for. But no, it has been impossible. They who want uniformly good photographic views, particularly of dark or overlighted workshops, must wait patiently in the hope of accident according that which the best chemicals and the most skilful handling cannot yet command. The time, no doubt, will come when it will be otherwise. It is, however, still apparently remote. Illustrating a book with photographs is as emblematical of patience as reading an American President's Message on a cold grindstone. Your thin-skinned "operator" is constantly being offended, your thick-skinned "operator" is as constantly offending, and your bungling "operator" has been tripped by an angle-iron, and his instrument is broken. Then there is the early awkward hour at which alone the sun will light some place sufficiently to allow the view to be taken; and the time, the long time that everything is said to be in shade. Next comes the trouble with the photographic printer, who hesitates to invest money in extra frames for the job, and has always an excuse ready for his neglect or shortcoming. To-day he will tell you that there has been no light for a whole week, and to-morrow his story is that the prints will lose colour and become worthless unless they lie eight-and-forty hours in water.

What are you to do? for after you have got rid of operator and photo-printer, there still remains the photo-mounter of the prints to be endured! A week, a fortnight, and a month pass, and you are exactly where you were. Meanwhile your publisher meets you in the street and reminds you that your book has been announced six weeks since in a half-page advertisement in the bi-monthly Circular, as well as entered in all the lists. Moreover, he adds there are inquiries for the book from Germany and other foreign parts, and that the questions of Dockyard Economy and Naval Power are at the moment before the public. The appeal is irresistible, and you yield to it, conscious that you have done your best, and that photography is not to be despised, although its results in November are occasionally less satisfactory than could be wished.

PHOTOGRAPHS.

FRONTISPIECE.—*Mersey Ironworks.*—Armour-plate Mill.

Millwall Company's Works, Millwall.
    PAGE
    Armour-plate Mill 221
    Interior View 224
    Interior View 226

Mersey Ironworks, Liverpool.
    Exterior View 232
    Exterior View 240

Messrs. Maudslay, Son, and Field's Works, Lambeth.
    The Foundry 261
    Engines of her Majesty's Ironclad Ship *Agincourt*, 1,350-horse power 261
    Erecting Shop 262
    Copenhagen Dock Engines 263
    Erecting Shop 264
    Erecting Shop 265

Messrs. Penn and Son's Works, Greenwich and Deptford.
    The Foundry 267
    Engines in Erecting Shop 267
    The Smiths' Shop 268
    The Scrap Forge 269
    The Heavy Turnery 270
    Deptford Pier Boiler Shop 270

Messrs. Rennie's Works, Deptford and Blackfriars.

    River View of Works 272

    Interior View 274

Mr. John Stewart's Blackwall Ironworks, Millwall.

    Exterior View 280

    Exterior View 281

    Exterior View 282

The National Boat-Building Company.

    Company's Works, East Greenwich 283

    Interior View 284

Mr. Charles Lungley's Deptford-green Dockyard.

    Interior View 294

    Interior View 295

Messrs. Samuda's Yard, Millwall.

    Interior View 296

Messrs. James Ash and Co.'s Yard, Cubitt Town, Millwall.

    Exterior View 299

    Interior View 299

    Interior View 300

DOCKYARD ECONOMY AND NAVAL POWER.

Chapter I.

THE DOCKYARDS.

Their present Appearance. The present appearance of one and all of the English dockyards is much the same. Deptford is from the Portsmouth pattern, and, in the main, between Keyham and Pembroke there is a close resemblance. When one is seen all are seen. Nor is this remarkable. The same heads conceived, the same money paid, and the same hands fashioned all the seven: Deptford, Woolwich, Chatham, Sheerness, Portsmouth, Devonport and Keyham, and Pembroke. All have building slips, although Sheerness has only one. All have docks, although Portsmouth and Keyham are the best provided. All are fortified, although the guns covering the sea and land approaches to Deptford and Woolwich are in store in the latter; although the Chatham lines are as ancient as 1710 and 1806; and Sheerness and Pembroke, as regards defensive works, are spectres by the side of Portsmouth and Devonport. The same London-Dock walls surround all, not that a foe may be kept out, but that common thieves may not break through and steal. The same London-Dock chief entrance has in all cases to be sought, before admission is obtained, and when it has been reached, the London Peeler uniforms occupy the wicket-stalls of the liveried officials of the establishment in Lower East Smithfield. Inside the chief dockyard entrance, the look of things is not more impressive than a glance into any barrack-yard where there are no soldiers. In front, or to the right or left, one or more infantry battalions might be drilled, if not officially inspected and reviewed. Not sailors, but military gentlemen, who are generally supposed to know nothing of the wants of sailors, laid out the great naval establishments which are the pride of so many Englishmen. Is it possible that they were actuated by ill-feeling to the navy, and had a leaning to their own sister arm, or that they participated in the dreamy notions of perpetual peace which took possession of so many people after Waterloo, and resolved on making the dockyards appear as dull, inoffensive, and picturesque as possible. Whatever the motive of those charged with the duty of laying out the dockyards, the judgment to be passed on them in their present form is, that they are better suited for ornament than use. They no more resemble the busy private shipyards (in one of which there is often more work done than in all the seven dockyards) than chalk resembles cheese.* A private shipyard is a hive of industry, with every foot of frontage occupied in building ships, and every foot of depth and breadth filled with shops, sheds, furnaces, and working places, so that at all times it is with difficulty, and occasionally with danger, that the visitor moves about. A dockyard, on the contrary, presents a *tout ensemble* of open ground and building, shrubs, trees, and official residences in which peers of the realm might spend their leisure. Everything unsightly, the vulgar outfit of the fleet, the hideous sawpits, and stacks of

---

* Of late the Thames Shipbuilding Company have been doing more new work than all the seven dockyards. The Millwall Shipbuilding Company will soon also be doing more.

timber, fill the water's edge, where it is no doubt presumed they remain unnoticed, and where of course an enemy would never think of harming them. Where it is necessary to carry on vulgar work, costly buildings have been erected, in which wretched labourers wear out soul and body for 13s. weekly and contingent superannuation. Where ships are to be seen in dock or basin there is the same straining at outside show; docks and basins too often containing showy rotten ships, which a single well-directed broadside would consign, with every one on board, to Davy's locker. The appearance of an English dockyard is therefore fine, and to some extent imposing, when set out with ships on which the paint-brush has been freely used. The money of a great country appears to have been spent to some purpose. Along the clean well-paved promenades, where starched Admirals, dim Commodores, pious Superintendents, sprightly young officers and others receive salutations and display their clothes or uniforms, a holiday hue adorns everything, as much as if the dockyards were established and maintained for the same reason as the fountains behind Nelson's monument in Trafalgar-square.

Their present State. The present state of the dockyards is in a certain sense implied in the appearance, but the distinction is apparent when appearance is supposed to mean no more than the outward aspect, while state is regarded as the uses to which the dockyards are applied. In the state of the dockyards there is the same identity as in the outward aspect. Deptford, Woolwich, Chatham, Sheerness, Portsmouth, Devonport and Keyham, and Pembroke are alike internally, unless in one or two respects of no importance. Keyham and Devonport, although joined by a short tunnel, are each possessed of dockyard and factory shops, in which skilled and unskilled men and boys are employed. Each of the other dockyards is similarly provided. In one and all of the seven, men and boys are at work from week to week and year to year storing masts and repairing masts, storing boats and repairing boats, making capstans and storing capstans, making sails and storing sails, making rope and rigging and storing rope and rigging, making oars, trenails, blocks, and storing them, mixing paints, repairing locks, hose, pumps, and other things; but Chatham Dockyard is alone graced with a lead-mill, a paint-mill, and a cement-mill. Pembroke, too, differs from all the others in being regarded as a mere building dockyard; and Sheerness also differs in being regarded as a mere fitting and repairing dockyard. Differences such as these are immaterial, but they deserve notice, inasmuch as they show the tendency to change that has from time to time agitated the official mind. When Sheerness was designated a mere fitting dockyard, and only provided with the means of building one ship at a time, it may be presumed that the thought of permanent ship-of-war construction in the private shipyards was seriously in contemplation; and on the other hand, when Pembroke was designated a mere building dockyard, it may be presumed that in the councils of the nation the reactionary party were again in power.

The Building Slips.* Step into one of the building slips in

---

* The following is the summary of the entire dockyard work in progress in the undermentioned dockyards at the time of my visit some months ago; reprinted from my reports in the *Morning Herald* and the *Standard*:—

Deptford.—Deptford is constructing the *Enterprise*, iron-cased sloop of 990 tons.

THE DOCKYARDS.

any of the seven dockyards, whether or no there are men at work, and a fair sample of the dockyard system in that one particular is obtained. The slip is covered

On the adjoining slip there is the *Favourite*, iron-cased corvette of 2,186 tons. In dock there is the *Salamander*, paddle sloop of 818 tons. On the *Enterprise* there are a few gangs of shipwrights; on the *Favourite* there is not a single shipwright; and the *Salamander*, an effete and rotten tub, has so many shipwrights at work that, even were they willing, they could not perform a fair day's work. Such is the whole present utility of Deptford Dockyard to the public, and the extent of the supervision of the officials.

Woolwich.—Woolwich is constructing the *Caledonia*, iron-cased ship of 4,125 tons. That is all the new work going on. In the docks there are the *Archer*, screw corvette of 973 tons, undergoing thorough repair; the *Alecto*, paddle sloop of 800 tons, undergoing thorough repair; the *Caradoc*, paddle vessel of 676 tons, undergoing thorough repair; and last of all the *Dee*, paddle store vessel of 704 tons, undergoing thorough repair. On No. 1 slip there is the *Dartmouth*, 36, screw frigate of 2,478 tons, in frame; on No. 2 slip there is the *Sylvia*, screw gun vessel of 695 tons, planked; No. 3 slip is full of timber; No. 4 slip is empty; on No. 5 slip there is the *Repulse*, 89, screw ship of 3,716 tons, with the outer skin; on No. 6 slip there is the *Wolverine*, 21, screw corvette of 1,702 tons. On not one of these slips is there a single shipwright at work; so that the extent of the ship construction and repairing of Woolwich Dockyard is the construction of the *Caledonia* and the repair of four small wooden craft, which it would have been wise economy to burn.

Chatham.—Chatham is constructing the *Achilles*, 30, iron screw vessel of 6,079 tons, and the *Royal Oak*, 34, iron cased-ship of 4,056 tons. This is all the new work going on, and there is no old work in progress. On No. 1 slip there is the *Salamis*, paddle despatch vessel of 835 tons; on No. 2 slip there is the *Reindeer*, 6, screw sloop of 950 tons, no deck; on No. 3 slip there is the *Myrmidon*, 4, screw gun vessel of 695 tons, advanced; on No. 4 slip there is the *Belvidera*, 51, screw frigate of 3,027 tons, in frame; on No. 5 slip there is the *Bulwark*, 89, screw ship of 3,716 tons, four-eighths; and on No. 6 slip there is the *Menai*, 21, screw corvette of 1,857 tons, in frame. On not one of these slips is there a single shipwright at work, so that the extent of the ship construction in Chatham Dockyard is the *Achilles* and the *Royal Oak*; the repairing *nil*.

Sheerness.—Sheerness is not constructing a single ship; and has one ship, the *North Star*, 22, screw corvette of 1,857 tons, in frame, with no work doing. In No. 1 dock, there is the rotten *Cossack*, 20, screw corvette of 1,296 tons, undergoing thorough repair; in No. 2, there is the rotten *Terrible*, 21, paddle frigate, undergoing thorough repair; in No. 4 dock, the gunboat *Cochin*, from the first division of the steam reserve, is undergoing examination; and in No. 5 dock, the rotten *Locust*, 1, paddle tug, is undergoing thorough repair. In No. 1 basin, the rotten *Erne*, gunboat, is undergoing thorough repair; in No. 2 basin, the rotten *Scylla*, 21, screw corvette, is undergoing thorough repair; and in No. 3 dock, the rotten *Vigilant*, 4, gun vessel, is undergoing thorough repair.

Portsmouth.—Portsmouth is constructing the *Royal Sovereign*, 3, iron-cased cupola ship, of 3,963 tons; and the *Royal Alfred*, 34, iron-cased ship, of 4,045 tons. This

in and capacious, and alike extravagant in its details and air of permanency. It is a structure for all time, based on the assumption that building slips will be always wanted in the dockyards, and that the form and capacity of ships of war are, and must always be, invariable. If there are men employed, they perform their task leisurely. If there are not men at work, it will be observed that the tool-chests, moulds, and materials are disposed of in faultless order; an essential part of the duty of a dockyard workman manifestly being, to give his first care to the tidiness of the place about him, that perhaps the sensibility of swell clerks, officials, and illustrious visitors may not be ruffled or their continuations ripped or soiled.

<small>The Storekeepers' Sheds.*</small> Step again into one of the storekeepers' sheds. This, let us say, is the reception-room for returned stores. Here all the

---

is the whole new work in progress. On No. 2 slip there is the *Dryad*, 51, screw frigate; on No. 3 slip, the *Harlequin*, 6, screw sloop of 950 tons; and on No. 4 slip, the *Helicon*, paddle despatch vessel; all in various stages of advancement, but with no work doing. In No. 1 dock there is the rotten *Esk*, 21, screw corvette, of 1,169 tons, undergoing thorough repair; and in No. 2 dock, the rotten *Curaçoa*, 31, screw frigate, of 1,571 tons, undergoing thorough repair.

Devonport.—Devonport is converting the *Ocean*, 34, screw, iron-cased ship, of 4,045 tons; and that is all the new work going forward. On No. 1 slip there is the *Bittern*, 4, screw sloop, of 669 tons, keel laid; No. 2 slip is empty; the next slip has been converted into a storehouse; on No. 4 slip there is the *Robust*, 89, screw ship, of 3,716 tons, partly planked; and on No. 5 slip there is the *Ister*, 36, screw frigate, of 1,321 tons, in frame; no work doing on any of the slips. In No. 2 dock, there is the rotten *Alert*, 17, screw sloop, of 751 tons, undergoing thorough repair; and in No. 4 dock, the *Minx*, water-tank vessel, and the *Tortoise*, lighter, are repairing.

Keyham.—In the North basin, the *Constance*, new, 51, screw frigate, is finishing; and the *Princess Royal* is changing boilers. In No. 2 dock, the rotten *Valorous*, 16, paddle frigate of 1,257 tons, is receiving thorough repair; and in No. 1 dock the rotten *Gladiator* waits the decision of the Admiralty, the dockyard officials being of opinion that the ship should be broken up.

* The storekeepers give bond for the proper fulfilment of their duties; but, being necessarily permitted to delegate some portion of their duties as regards the issuing of stores to their clerks and storehousemen, are not wholly responsible for the safe

moveables of all the ships paid off at this particular dockyard are thrown down violently on the floor and sometimes smashed. Let us pause a moment as the crowded lighters are approaching, and the use of the reception-room will receive demonstration. The two lighters are at length made fast and the crowd of sailors and marines jump ashore and reach the quay. To all appearance the sailors and marines have not washed for some time, nor have their clothes been brushed. Inquiry, however, sets the mind right on these points, for the crew were as clean as pins some hours ago. Since then, to use their own expressive words, they have been tearing the guts out of the old brute. They have been in the inside of lockers, store-rooms, and what are familiarly termed bunks, in which there is neither light nor wholesome air, but by some unaccountable means plenty of dust, waste, and mildew. The consequence is that the prim sailor and marine becomes as clammy as a viewer of City sewers, or as dusty as those whose occupation is to knock down aged buildings within the precincts of the Middle Temple. The discharge of the two lighters now begins. Several sailors and marines take armfuls of old rope, others shoulder rope yarns, some carry blocks, half a dozen drag out chain, a few stagger with the weight of ponderous tackles, but the great majority frisk about with not very exemplary loads of old iron, wrenched by sheer strength from the interior of the paid-off ship. Into the reception-room the crowd presses, down they cast their burdens, and back they turn to the lighters.

---

custody of the property under charge.—Page 590, Appendix to Report of Commissioners; 1861.

The committee recommend greater control to be exercised over the stock of stores in the factory storehouses by the storekeeper of the yard, and over the cash payments of wages of the factories by the accountant of the yard.—Page 92, Report of the 1859 Committee.

The work is devilish, and because it is devilish it seems to be enjoyed. When the sailors and marines with the lighters have finally disappeared in the offing, the dockyard labourers begin to amuse themselves, sorting the ruinous heap of everything and carrying the different articles to near or distant stores, from which they will again be carried and abused when the next ship is commissioned in the dockyard.

The Workshops.* Step last of all into one of the workshops. The sight is a strange one after a previous visit to a shipyard workshop. A shipyard workshop is as well stuffed as a common lodging-house in St. Giles's. Planing, drilling, slotting, punching, clipping, and other machines take the place of shake-downs and bedsteads, and greasy horny-handed mechanics the place of tramps, illegitimate shoeblacks, and Covent-garden sneaks. In summer the air is close and balmy with the odour of the crudest oil. In winter a red-hot stove heats to excess the immediate neighbourhood, but leaves unwarmed the distant parts; or space cannot possibly be found for a stove, however willing the firm may be to supply one and provide the coal. In a word, a private shipyard workshop is always too small by half, and when an enlargement takes place it is soon usually found to be as much wanting in length and breadth as before. In a dockyard workshop, on the contrary, one is less likely to be

---

* There is an old custom in working the copper and brass mills to limit the men to a given stint of work; while the conductor recently appointed from private trade considers that the men ought to do about one-fifth more than this for their present day's wages.—Page 61, Report of the 1859 Committee.

The committee found the arrangements for the supervision of the men while at work in the various workshops to be generally objectionable. The offices of the foremen, inspectors, and leading men were not such as the men could be seen from, being in many cases detached from the workshops, and in others without any window looking into the workshops.

inconvenienced by hamper or lost among machines than to be lost in space and worn out by walking over ten times—nay, in Keyham a thousand times—the house room that is required for all the work and workmen. Magnificence as regards space, and, as far as decency will permit, as regards structure, is the great idea of the dockyard workshops. Bigger workshops than those of the private shipyards, or than those of France, and, if possible, than those of all foreign countries together, has always been a weakness of the Admiralty. Surely more contemptible illiberality cannot be conceived. A dockyard workshop resembles nothing so much as Westminster Hall with a dining-table and a dozen chairs in the centre at which half a dozen Queen's Bench witnesses are taking things comfortably. Straggling machines fill out the walls of the dockyard workshop, and the centre is available for a lecture or a dance. The great Keyham factory workshop covers as much ground as would form a moderate Alexandra Park, and here and there a half or whole acre is partitioned off for shops of some kind, while the major portion of the factory lies fallow. All the wasteful factory votes of many years have not sufficed to make much more impression on the great establishment at Keyham than all the shiploads of starved cotton-spinners and other emigrants in the colonisation of Queensland and Victoria. Such is the state and appearance of the English dockyards; a state and appearance disgraceful to the Admiralty and Parliament, and unjust to the taxpayers of a heavily burdened country.

The first intention of the Dockyards. Having now conveyed a pretty accurate impression of the present economy of the dockyards, let us next turn briefly to the past. The dockyards were originally established for what were deemed good reasons of State policy. Without going as far back as the time of Henry VIII., it will suffice to state that when British cruisers appeared off Toulon, Copenhagen, or New England, or when French cruisers appeared off Portsmouth, the means were to be at hand for repelling them; when British, French, American, or Russian cruisers returned to port disabled, facilities were to exist for refit or repair; and in peace and war the dockyards were to furnish the new additions to the fleet. Such was the first intention of the dockyards. They had their origin in the necessities of naval war and in the poverty of the industry of the time. They were intended to provide that which could not be conveniently or so well provided elsewhere. At that period England, Continental Europe, and America present a striking contrast to their state just now. The mechanical and manufacturing arts may be said to have been since created. Since the establishment of the dockyards, modern shipbuilding takes its rise. The shipbuilding that then and that long subsequently existed was of unsightly small craft, rude, ill-finished copies of the old school that had long preceded, and it was in the hands of quaint old fellows who were particular as to the season when keels were laid, timbers set up, and the first strake of planking fastened. However, to do them justice, shipbuilding was not encouraged greatly, and they were not behind their neighbours in general intelligence and skill. Commerce then flowed within narrow bounds. The impression then existed that it was ruinous to receive commodities from other nations, and only profitable to bestow them. Only between a country and its colonies was trade unshackled, and such a trade could at the best furnish an intermittent and feeble impulse to the building of sea-going ships from year to

THE DOCKYARDS. 11

year. So when it became necessary to possess cruisers, squadrons, and fleets, Lord High Admirals had no choice but that of building for themselves. Contracts they could not give, because there were no private firms to take them. Left, therefore, to determine between building ships and not possessing ships, they necessarily adopted the former course.

<small>Consequence of this first intention.</small> There is an obvious consequence of some importance arising out of this first intention. It is that as the dockyards had their origin in the necessities of naval war, and in the absence of great private establishments in which ships of war could be built, there is not another word to be said in support of the continuance of the dockyards than that they exist, or, what is the same thing, that we possess them. Manifestly, if up to the present time war had been unknown, and all at once its ghastly forms presented themselves to men's minds as something with which at last they must become familiar, no sane man would have thought of providing ships in any other way for the public service than ships are provided for private persons, because if it were impossible to supply the public service properly, the question would arise whether it would not be as much the duty of the Government to save private persons from what we shall call the rapacity and incompetency of the shipbuilders as to save the public. From this conclusion there is no logical escape. Some may answer that the dockyard work is better and more to be depended on than shipyard work; but surely it must be apparent that this is only saying the dockyard officials are incapable of superintending shipyard work, and that workmen cannot perform their task in so satisfactory a manner outside of

the dockyards as they can within. The objection, it will be shown hereafter, is altogether groundless. The first intention of the dockyards was to provide that which could not be conveniently or so well provided elsewhere; therefore all that the warmest friends of the dockyards can rationally say in support of them is that we have them, and for that reason we ought to use and cherish them. But exactly the same logic would lead us to use and cherish the coats of mail in the Tower of London, because we have them; and the battle-axes, broadswords, and unmelted cannon of the last century that still encumber our great arsenals.

<small>One source of mystification.</small> Why this consequence of the first intention of the dockyards is lost sight of in all the discussions on the dockyards admits of being explained. From long-formed habits of association with forts and lines of defence, the fact has come to be overlooked that any coast harbour in which ships of war may be built, fitted, refitted or overhauled, is to all intents and purposes as much a dockyard as Portsmouth or Toulon, although the only ditches in the neighbourhood are those for draining land, and the only guns the farmers' fowling-pieces and the rifles of the Volunteers. We have come to invest dockyards with great defensive attributes. They are regarded as the sole refuge for our ships of war when great expeditions are being planned, or when our ships of war cannot keep the sea in the presence of an enemy. Sometimes we are even told that the dockyards are the backbone of all naval power. But what in reality are the advantages of defended dockyards? If they might repel an audacious enemy, would they not constantly invite attack? If they might provide shelter for our ships, and afford

facilities for combinations into squadrons and fleets, would not all the facilities and shelter ever likely to be required be found without outlay or preparation elsewhere? We may be well assured that the issue of a war between two naval Powers must depend less on the strength of a few points of coast than on the multiplicity of undefended points of coast where the construction and repair of ships of war might be carried on. England may have its Chatham, its Portsmouth, and Keyham, and France its Cherbourg, Brest, and Toulon, but, manifestly, the power of naval resistance and aggression extends far beyond these strongholds. The strength of England in well-defended dockyards is really contemptible by the side of the strength of England in undefended shipyards. From the Pentland Frith southward round the English coast to the Caledonian Canal, and from Belfast Loch southward round the Irish coast to Belfast Loch again, there is, irrespective of the dockyards, as much available shipyard resource in the three kingdoms as there is in all Continental Europe.

Confirmation of the value of undefended Dockyards. The American war of 1812 furnishes abundant proof of the value of undefended dockyards. The supposed naval weakness of the United States was the chief occasion of that war, and yet there is no naval war on which England ever entered that reflects less honour on its arms. The Americans built and sent ships to sea in spite of us, and many of those ships overpowered ours. Every considerable commercial harbour in the United States became a dockyard, from which ships of war issued, and towards those harbours disabled American ships bore up indifferently, with the assurance that repair and refit could always be obtained. And when our ships followed them, or appeared before one of those commercial harbours, there was nothing practically to destroy. In none of them were there great accumulations of public stores, and consequently no effective blow could be ever struck at our enemy. England could only fret at the unexpected and bitter check; and the invention was a happy although transparent one, that America had stolen a march on us in the construction of more heavily armed ships of the same classes. The acknowledgment of the truth would not have answered, that why America succeeded and England failed was solely owing to the fact that England had formed its estimate of naval strength by the possession of great dockyards, while America attached no importance to great dockyards, but made the most of its rude harbours, its forests, and our own seamen. It is not yet too late for England to profit by the dearly bought experience of the last war with the United States.

<small>The defended Dockyard theory.</small> The defended dockyard theory is, as just observed, that as long as a nation possesses strong dockyards it can do pretty much as it pleases. This is a short but fair statement of the theory, and obviously its weak point is that now-a-days there are in reality no strong dockyards. Improvements in gunnery and the iron-casing of ships divest the strongest dockyards of more than half their once boasted power. Frenchmen now smile at Spithead, and at the new covering works in progress, and Englishmen in turn smile at Toulon. The science of attack always has and always will keep pace with the science of defence, because practically that which is good for one is so for both. Naval architects and engineers may now be found who will engage to build ships that would carry granite

fortresses on their decks, and mount guns in such fortresses of larger calibre than artillerists would dare to fire—unless from the security of the splinter-proofs at Shoeburyness. Then if Portsmouth by any chance could not be destroyed by bombardment from the sea, an army establishing itself in force in its rear would very soon enforce submission. So with Toulon, and so with all the great dockyards. Repeated failures at reduction by bombardment at Vicksburg and Charleston prove no more than that ironclads of the Federal pattern were long unequal to the work. They leave untouched the question of the power of the ironclads of Europe in such enterprises, and the equally important question of the invincibility of European armies in the field when supporting such attacks. Thus the dockyard theory does not hold water. It is a theory the basis of which is an impossibility, and yet, strange to say, possibility is boldly assumed in its support. It is assumed that Portsmouth, Toulon, and Cronstadt would keep out an enemy however powerful, and the assumption granted, it follows that a nation possessing defended dockyards has the means of constructing, fitting, refitting, overhauling, victualling, and concentrating fleets at pleasure.

<small>Dangerous tendency of the theory.</small> The theory of defended dockyards has one most dangerous tendency. It fosters a strong dependence in mere outlays, although those outlays may be of no real value. Money is expended in enormous sums, and people as a consequence come to think much the same thing of the defended dockyards as the Austrians thought of their pampered and overbearing legions before Solferino, and the demoralisation that followed the overthrow of the hollow Austrian system is to be feared, even in this country,

were our defended dockyards to be demolished by an enemy at the beginning or in the course of a great naval war. Austria hoped everything of its soldiery, with a liberal hand spent its means on them, and when they failed the country was broken in spirit and without resource. True, Englishmen possess qualities the absence of which has often, and may many times again, render Germans contemptible in the eyes of Europe; but nevertheless, were Englishmen to find that they had been trusting to a reed in the matter of defence, the effort would not be an easy one to forget the defended dockyards which had before been lauded, and confide in the shipyards, the harbours, and the open roadsteads of the kingdom, which before had been lightly thought of. England suffered deep humiliation from the relatively unimportant break-down of the Horse Guards' pipeclay and leather stocks in the Crimea. God forbid that too great dependence should now be placed in the defended dockyards!

The cognate Trades. But returning to the subject. With shipbuilding set a-going in the dockyards, the establishment of the cognate trades followed. Ships of war could not be built in a given time nor at even a remote approach to a certain sum, unless everything entering into the construction was at hand. At that period the outlying districts on the coast communicated at best with the centres of industry and population by bullock-cart and team. There were no telegraphs, and no canals nor steamboats. Carrying coal from Newcastle was unheard of, and unthought of. Land transit in those days from Newcastle to Portsmouth or Plymouth would have invested coal and other bulky articles with a value that would have restrained even dockyard buyers.* So England, after taking to the building of ships of war in the dockyards, was from necessity obliged to advance a step further and as far as practicable produce on the spot all that the construction of ships of war required. Shops and sheds were erected, that masts, boats, sails, rope, rigging, and other things might be procured, and it is a suggestive and startling fact that those long-since superfluous adjuncts should still survive. That what was requisite in the construction of ships of war in Henry VIII.'s days should still be kept alive by annual Parliamentary votes in the year of grace 1863-4, is perhaps creditable to our respect for old things, but eminently discreditable to our common sense. That the cognate shipbuilding trades should still exist in the dockyards is a proof, were one wanting, that since the establishment of the dockyards they have not yet been brought into anything like harmony with the times.

<small>The growth of the Dockyard Towns.</small> Contemporaneous with the growth of the dockyards has been that of the towns clustering round them. The history of these towns is still unwritten. With the French dockyard towns it has been much the same as with those of England, while with those of Russia and the United States what is true of the French and English is true of them in only a small degree. From the first Russia has been a despotism, and corruption in the dockyard towns has not been necessary. In America the ballot-box and the other safeguards of an advanced state of freedom—often abused, no doubt—have been sufficient to preserve the

---

* I do not think it necessary to repeat what I have said elsewhere on the subject of transit.—See *American and Indian Transit:* Trübner and Co., 1859.

dockyard towns at elections from sale and purchase; but proof is not wanting that both in America and Russia jobbery and intimidation have always to some extent prevailed. A Russian Admiral and an American senior naval officer are persons worth pleasing, and that they are pleased goes to some extent to show that dockyards and corruption are inseparable. Between the corruption of the French and English dockyard towns there is not much to choose, but the latter no doubt bear the palm.

<small>The position of the Dockyard Towns.</small> The dependence of the dockyard towns, as long as dockyards are maintained, is, however, in a certain sense necessary. Round the most faultlessly conducted dockyards towns of more or less dependence must in course of time be formed, and if the Government of the day were not to turn the dockyard towns to some account they would be wrong. Still, linked as the dockyard towns are with the dockyards, the consideration to be kept in view stands out clear enough. Up to the point that the connection between the towns and dockyards operates beneficially to the public it is harmless, but when the not at all difficult to be defined limit has been passed it of course becomes an evil. For example, in justification of the construction of Mr. E. J. Reed's sloops of war in Deptford Dockyard, it has been said, although not by Mr. Reed, that the shipwrights who otherwise would have been idle are employed, to the great benefit of Deptford. Obviously here is a glaring disregard of sound principle. Then again, when it is said, in justification of the continued construction of ships of war with wooden bodies or wooden bottoms, that the excessive stock of timber at the present time on hand will be consumed, the trans-

THE DOCKYARDS.

gression is alike apparent. And to take an instance of another kind, when in the dockyards an old-fashioned shipwright, a man with vulgar prejudices in favour of timber and against iron, is raised by favouritism or improperly in any other way to a high position, and in that high position is called on to express a professional opinion which shall guide the Admiralty either in the construction of wood or iron ships, the abuse is not less patent. Whenever, therefore, dockyards come to be regarded as a means of providing employment to a resident population, as a means of working up material that happens to be on hand, or as means of employing improper persons, the dockyards are at once degraded to the condition of establishments for the relief of tradesmen and able-bodied workpeople.

No hardship in this judgment. In this judgment there is no implied hardship to any class. Let it be assumed the rule were laid down that unless the construction and maintenance of ships of war could be obtained on as advantageous terms in the dockyards as out of them, construction and maintenance in the dockyards should cease. What, then, would be the position of the dockyard operatives, and others in the dockyard towns? The services of the operatives and the capital of the tradespeople would cease to be required, but, as a set-off, the services of the one and the capital of the other would be wanted wherever the construction and maintenance of ships of war were transferred. Shipwrights, caulkers, and others would cease to be in request at Deptford, Woolwich, Chatham, Sheerness, Portsmouth, Devonport, and Keyham and Pembroke, but they would be wanted on the Thames, the Mersey, the Clyde, and the Tyne. So it would be, too, with the tradespeople of the dockyard towns. The only necessary sacrifice would be the very acceptable one of change. For the public good, labour and capital would undergo a transfer—nothing more. But, it will be urged, the owners of property would be ruined by the exodus of operatives and tradespeople. There is force in the objection, but when the Corn Laws were repealed the British farmer was not only laughed at but very properly abused for robbing society during many years. Who, again, sympathised with the shipowners in their sore bereavement when the British carrying trade was thrown open to the world for the benefit of the British public? Is there, then, any wrong implied in dealing with the paltry, corrupt dockyard towns* as we have dealt before with great national interests? Why should society be taxed for the benefit of a handful of landlords in the dockyard towns, who, in too many cases, have only provided wretched hovels for the shelter of the labourer and mechanic? The dockyard towns in the main are the filthiest towns in England. Built with the possibility of bombardment staring landlords in the face, and continued to the present time with that risk counted on by the landlords and paid for by the tenants, their complete extinction should occasion no regret; the returns for vested capital and the comfort of the working classes both considered.

<small>The Dockyards not wholly unproductive.</small> That the dockyards have not, however, been wholly unproductive is not for a moment to be doubted. They have given the taxpayer something for his money, although it must be confessed they have given him very little. With them

---

* 12th. That the officers and men in the dockyards should be placed under the same restrictions as to voting at elections as the officers of the Post-office, the Customs, and the Inland Revenue. Page 7, Report of the Commissioners; 1861.

as they are, glorious naval actions have been fought that more than compensated for the relative weakness of the army, and gave to our arms the prestige that they still retain. With the ships of war that our dockyards turned out, a long line of naval heroes conquered. This is a proof that the treasure spent from first to last has not been altogether thrown away. But it leaves the question where it was, whether a dockyard system suited for one period is suited to all others? It is long since a Nelson fell on the deck of the *Victory*, and every year bears witness to economic changes as well as developments and applications of science, directly and indirectly bearing on the construction and equipment of ships of war. To the dockyards, however, we owe at least one grudge. When dockyards were first established an attempt at divided labour would have been unwise, if not futile. No country then possessed a supply of men trained in the minute details of industry so that all the parts of a ship of war might be fashioned and put together at the minimum of cost, and consequently no choice remained but the simple and costly one of filling the dockyards with a ready class of workmen who were to put their hands to everything. And the creation of separate classes of workmen would at the period have proved no great boon to the State or the workmen. For, think as we may of the fits and starts of naval armament of which the public have of late years so much complained, our fathers were prone on all occasions when the toil of war ceased to lay aside the trappings and the arms with which they and the State had become encumbered. They were willing to support the dockyards when national existence might have seemed to depend on them, or when advantages more or less solid were in anticipation; but when the maintenance of the dockyards could only be justified on the plea that to prevent war people must be ready to engage in it, their choice as a rule was to enjoy without diminution the blessings that peace conferred, and leave the future with its hypothetical fighting to itself. Peace, therefore, implied the discharge of dockyard workmen, and manifestly if those workmen had been trained to one branch only of their occupation they would have suffered less or more from the circumstance. Obviously, also, the State, then relatively weak and poor, would not have gained by a course that terminated in such a manner. Hence a vicious working system was early fostered in the dockyards. The principle of the system is that the division of labour is inimical to the privileges of the dockyard workmen, who on all occasions give their votes to the nominees of the Government in power or of those who soon expect to be in power. Dockyard workmen claim a prescriptive right of doing in the main as they please. The men who convert the timbers of a ship are in their own opinion fit and proper workmen to raise the timbers, to plank them, to lay the decks, to erect the interior fittings, and do all the odds and ends of jobs that fall in their way. So there is practically no division of labour in the dockyards. A dockyard apprentice becomes a Jack of all trades, and if a pushing lucky fellow, a fellow particularly who has done good service on a pinch in beating up recruits for the polling-booth, he may eventually be promoted to the rank of carpenter, draftsman, master shipwright, foreman engineer, naval constructor, master blacksmith, or probably to the post of chief engineer and inspector of machinery. The only two posts that dockyard apprentices and labourers may not aspire to fill are those of chaplain and staff surgeon. For all the others in the patronage of the Admiralty or the hangers-on of present and prospective lords, they are held qualified on the ground that ample opportunity has been afforded them of getting crammed sufficiently for any one of the various dockyard callings.

Dockyard administration.* Last of all, let us turn to administration. Dockyard administration is in a sorry and almost hopeless plight. The hopelessness arises from the complete impotency for reform of Royal Commissions and Parliamentary or Admiralty committees. Admiralty committees have been appointed to assuage the indignation of an ill-used public, and Parliamentary committees and Royal Commissioners have followed when the ill-used public have found out the

---

* We regret to state that in our opinion the control and management of the dockyards is inefficient. We are of opinion that the inefficiency may be attributed to the following causes:—1. The constitution of the Board of Admiralty. 2. The defective organisation of the subordinate departments. 3. The want of clear and well-defined responsibility. 4. The absence of any means, both now and in times past, of effectually checking expenditure from the want of accurate accounts.

As a proof of the inaccuracy of the detailed accounts, we refer to the report of an investigation into the accounts at Woolwich Dockyard. The examiners report that they have discovered 7,906 errors in the accounts from April 1 to November 30, 1860, of which 6,566 are in the rating, valuing, and totalling of timber and store notes in the Storekeeper's department, varying in amount from 1d. to £490; 208 in the rating, valuing, totalling, and proving of workmanship notes in the Accountant's department, varying in amount from 1d. to £11; 874 in the postings totalling them, and preparing returns Nos. 88 and 89 in the Accountant's department, varying in amount from 1d. to £400; 71 in the postings to aid totalling of factory portion of returns Nos. 88 and 89 in the Chief Engineer's department, varying in amount from 1d. to £4 9s.; and 187 in the postings and totalling of return No. 8 in the Storekeeper's department, varying in amount from 1s. 6d. to £365 9s. It appears from the same document that a sum of £4,480, for engines and boilers of the *Ranger*, was omitted to be entered into the monthly return for May; and that a sum from £1,000 to £1,200 per annum, for the time of certain workmen, was twice charged from the 1st April, 1858.

The system of accounts is elaborate and minute, but, so far as we can judge, its results are not to be relied upon for any practical purpose.—Report of the Commissioners; 1861.

trick practised on them. But Royal Commissioners and Parliamentary committees have always shared the fate of Admiralty committees: that is to say, their reports have been as contemptuously pigeon-holed and forgotten. As regards general dockyard and general Admiralty affairs, the issue of a Royal Commission or the appointment of a committee of any kind means neither more nor less than the adjournment or shelving of the question until again, in the course of time, the old outcry is renewed, but only to be again met with a fresh inquiry. So the circumlocution wheel turns round, the dockyards and the dockyard system receive a new lease of life, and system and expenditure proceed unchanged. During the last session of Parliament another Committee of Inquiry was proposed, but owing to the lateness of the motion it was withdrawn on the understanding that when Parliament meets again it will be brought forward. Had the motion stood in the name of Lord Clarence Paget or Mr. Stansfeld, its meaning would have been intelligible, but in the name of an unofficial member it suggests only the probability of the dockyard system of administration remaining long the blot and scandal that it is.

The testimony before the last Commissioners.* The fluctuating and unsatisfactory character of the administration of the dockyards was, and certainly not for the first time, plainly testified before the last Royal Commissioners. Sir A. Bromley, 775; Sir Baldwin Walker, 472, 473, and 1,133; Col. Greene, 4,898 to 4,900, stated that, collectively and individually, the management of the dockyards has been of a fluctuating and sometimes of a

---

* Report of the Commissioners; 1861.

contradictory character. Sir Baldwin Walker, 547, 596, stated that in matters relating to his department several different Lords had specific duties of superintendence. In other words, the Controller of the Navy served under a Babel of authority; Lord Buttons ordering buttons, and Lord Hooks-and-Eyes ordering hooks and eyes. Sir Baldwin Walker, 238, 598, further stated that any two Lords and a secretary form a Board of Admiralty —in other words, the First Lord, the Civilian Lord, and Lord Clarence Paget might hold a Board meeting and issue Board orders or instructions from the Trafalgar at Greenwich over a whitebait dinner; while on the same day the other secretary and two more Lords might hold a Board meeting in the House of Commons refreshment-room, in one of the waiting-rooms at the Reform Club, or in any West-end chop-house, counter-ordering and counter-instructing—in a word, checkmating—the other Board of Admiralty at Greenwich. No doubt good sense and politeness prevent many unpleasant jars, considerable inconvenience, and save the public large sums; but mistakes, of course, will and do happen. The expectation that absent colleagues will always sanction street-corner deliberations of the Board of Admiralty—two Lords and one of the two secretaries being present—must necessarily be sometimes disappointed, and the humiliation suffered of rescinding orders and instructions before the ink has dried upon them. Sir Baldwin Walker, 472, further stated that during the twelve years from 1848 to 1860, the office of First Lord of the Admiralty was filled by seven different persons —namely, Lord Auckland, Sir Francis Baring, the Duke of Northumberland, Sir James Graham, Sir Charles Wood, Sir John Pakington, and the Duke of Somerset. In other words, the administration of the dockyards goes into apprentice hands once every eighteen months on the average.

<small>Administrative cobbling.</small> Mr. Torren's *Life and Times of Sir James Graham** furnishes a complete and instructive compend of what may be called administrative cobbling, and a plain statement of the constitutional fears of the right honourable baronet. Sir James Graham, in abolishing the Navy and the Victualling Boards in 1833 to secure increased concentration and responsibility, stopped short at the complete concentration and responsibility implied in the appointment of a Secretary of State or Minister of Marine for the Navy, because† " who could tell what limitations might be introduced into the terms of that commission, or how long it would be before the department with all its vast patronage and influence was conferred on a Prince of the Blood? When reconstituting the War-office in 1854 it had been at first understood that the Secretary of War was to have the supreme control of the army; but when the commission came to be framed difficulties arose, and the result had been that the Horse Guards still exercised co-ordinate authority without responsibility to Parliament." The reason is one of immense gravity and calls for a disclaimer on the part of the Crown. Surely the offset to administrative reform at the Admiralty is not the installation of Prince Alfred as Lord High Admiral. If so, little penetration is required to foresee that the public will never sanction such a compact. Attached as all classes of Englishmen are to the Royal Family, he would be a bold man who ventured to affirm that indifference to the encroachments of the Crown is a condi-

---

* Saunders, Otley, and Co.; 1863.
† Page 654, vol. ii.

tion or consequence of genuine loyalty. Nothing, on the contrary, is more patent than that the high esteem in which the Royal Family is universally held is in a great measure owing to the practical abandonment of the patronage and executive functions of the Crown to the Government of the day. We esteem our Queen, as much as anything, because she is regarded as self-sacrificing and most loyally disposed to acquiesce in all that is calculated to promote the public welfare. We love our Queen because she is thought incapable of seeking the sordid advancement of her family and house. When Prince Alfred is seen in his lieutenant's uniform we are pleased to think that at least some portion of his time is suitably employed, and no one certainly dreams of one in training for the irresponsible expenditure of £12,000,000 annually. If the Queen is wise she will reject the counsel of Court flatterers and either expunge Prince Alfred's name from the Navy List or abandon the thought of his ever filling a position from which public opinion would assuredly compel him to retire. The Princes should not be more obtrusive than their worthy father, and will best consult their happiness by advancing art and science. The Victualling Board and the Navy Board which Sir James Graham got rid of were constituted, like the Board of Admiralty, by patent from the Crown. The Navy Board prepared its estimates for shipbuilding and repairs, and over the expenditure the Board of Admiralty had no control; the control of the Admiralty extending no further than the ordering of ships and repairs. The Victualling Board was in precisely the same position, and the case made out against them was that, as regards the Navy Board, in four years £1,029,000 less than Parliament had voted had been expended on timber and material and applied to purposes which Parliament had not approved; and as regards the Victualling Board, that hospitals and other buildings had been constructed in the same loose and improper manner.

The Board inspections. The Board inspections are one of the great points in dockyard administration. When the Board have got through the drudgery of the Session, the character of inspectors is assumed during the first month of the recess, and no doubt the newspaper accounts of the reception of Lord Buttons and Lord Hooks-and-Eyes at Portsmouth have conveyed to many the impression that after all dockyard matters are not so bad as they are said to be. Let us, however, attempt to place the Board inspections in their true light. Buttons, Hooks-and-Eyes, and one secretary meet by appointment at Waterloo, and, the customary greetings over, away starts a special train with the three worthies. This is a £5, £10, or £20 waste to begin with. On the arrival platform at Portsmouth a guard of honour may or may not be dispensed with; but at the dockyard gates a strong body of marines is under arms. The Board inspection was intimated a week or fortnight in advance; or should the inspection be a private one and a surprise, two or three days' underground or backstairs warning will only have been given. The Captain-Superintendent is in full feather, and so is every one with whom Buttons, Hooks-and-Eyes, and the secretary are to come in contact. Should it be desired, drums beat, fifes play, and there is a display that might embarrass Royalty. If, however, the "Lords" are quiet, they will seek no more than a turn-out resembling that which takes place at a parish beating of the bounds: the master shipwright and assistants, the storekeeper and assistants, the accountant, chief engineer, store receiver, superintending civil engineer, and their assistants, forming the obsequious and smiling retinue. The storekeeper is of opinion that the chief object of interest to "My Lords" is a strangely shaped anchor brought up by the dredger the other day. My Lords assent. An hour's pleasant chat follows, and the antiquity of the anchor is learnedly determined by the secretary. The chief engineer has just returned from Birmingham, and, if it is the pleasure of "their Lordships," he would submit a scheme of building ships with scarcely any labour and at very little cost. One Lord nods, the other shakes his head, and the secretary turns his back: the noes, therefore, have it, and the chief engineer is silent. The master shipwright, if it is the pleasure of "their Lordships," would invite their attention to a frigate at present on the stocks and which was ready for launching two years ago. Their Lordships proceed with the master shipwright, and gratify him with half an hour's earnest abuse of all the shipbuilding plans before the public. He is told that the Board will give that consideration to his professional opinions to which they are entitled. The storekeeper has then something to say about the supply of coal to the policemen; but the secretary, anticipating the infirmity of their Lordships, suggests luncheon. Their Lordships thereupon adjourn to the official residence of the Captain-Superintendent, and from the official residence a second adjournment takes place to the well-known hotel, at which the health of the Captain-Superintendent is first proposed with all the honours, and afterwards the health of the individual members of the Board of Admiralty is disposed of in the same friendly manner. Should the Board inspection occupy two days, there are, of course, two days' expenses chargeable to contingencies.

<small>Co-operation Morning Meetings.*</small> The co-operation morning meetings of the principal dockyard officers are one of the reserve cards of the Admiralty when sore pressed on the subject of dockyard management. These meetings were originally intended, and are still intended, to bring the heads of departments into harmony, so as to obviate the possibility of imitative duality; the engineer and civil engineer, or the shipwright and the storekeeper claiming in turn to be supreme. The idea is a good one, but scarcely applicable to officials who hold their post by so sure a tenure. Were the dockyard officials, like the foremen of the private shipyards, liable to be paid off for any shortcoming, the morning co-operation meetings would answer admirably. But why should morning co-operation meetings be anything but a bore to gentlemen looking forward with the full assurance of hope to superannuation? What to them is economic harmony, or wasteful discord? Their bread-and-butter is secure, although for a whole year or more not a ship is built, fitted, or repaired in the dockyard where they are. They, therefore, take the morning meetings calmly; in some dockyards they have even been discontinued altogether. Let us attend one of these gatherings. The last man to arrive is the master shipwright, and he is so late that he might as well have remained away. His excuse is that he was out last night, and that soda-

---

* The committee have found at some of the yards that the morning meetings of the officers with the superintendent at his office have not been taken advantage of to the full extent that was contemplated when they were instituted, and that they have fallen into a mere reading of the orders of the day, occupying a few minutes only.— Page 2, Report of the 1859 Committee.

water with something in it would now do him good.
Not a word passes about the dockyard work in progress
or in contemplation. Another morning the master
shipwright is the first man, and he paces the room
impatiently. He has a communication to make. Before
he begins one of the co-operators breaks out in a hoarse
laugh: he is aware what is coming. Yes, in one of their
moments of despondency the Board—the damned Board
—have been tampering with the inalienable rights of
the dockyard officers. Lord Buttons has pitchforked
an uncircumcised stranger not only into the service but
actually over the heads of men old enough to be his
father. The master shipwright wishes Lord Buttons
and the Admiralty at the devil. The engineer smiles.
The carpenter recollects several other cases of the same
kind in the course of his long experience. Some other
person has often wondered whether, in the case of an
infringement of the rights of the dockyard officers, a
strike should not be resorted to as a means of bringing
the ungrateful Board to a sense of their duty, and the
country to a proper appreciation of the great services
rendered by the dockyard officers. Another one details
the pedigree of Buttons, whose father was a Leather-
lane oyster-seller, and denounces him as an interfering,
incompetent upstart. These conversations having occupied the full regulation period for the co-operation morning meetings, the master shipwright bids his chums good
morning. But on some occasions there is real business
talked of. The corvette *Witch* is in the basin, and the
question is, whether dry rot has proceeded far enough
to justify what in effect would be the complete rebuilding of the ship. The carpenter urges the rebuilding
strongly because there are a great many shipwrights
just now out of work, and it would be charity to find a

job for them. The storekeeper takes the same side, because his hands are full of all kinds of useless stuff. The engineer will be happy if the rebuilding takes place, because he wants to see water in the new dock, and its capabilities displayed. The civil engineer is of opinion that the rebuilding would afford a very good trial of his last invention, which, it will be remembered, is to tear out one or more timbers from the frame of a ship without disturbing those to which it is joined, or without shaking the structure in any perceptible degree. The smith says a great many of his men are in want of more to do, and the state of the corvette is such that a very considerable amount of blacksmithing will be required. The master shipwright thereupon declares that the ayes have it, and a messenger is forthwith despatched to the Controller's office with a submission for the reconstruction of the corvette, signed by all the members of the co-operation meeting. To this array of professional authority the Controller bows, My Lords also bow, and another wooden tub is restored to the first division of the steam reserve at an aggregate outlay that would have sufficed to build a new ironclad.

<small>Piety and uncharitableness.</small> As a matter of course, piety and uncharitableness go hand in hand in the dockyards. In those dockyards over which a crack-brained naval officer has been placed, who is painfully alive to the sin of keeping the Chatham Dockyard metal-mills at work on Sundays,* as they always are,

---

* The mills are closed from ten on Saturday till six on Monday morning, so far as the manufacture is concerned, and in the meantime the repairs are done.—Evidence of Mr. Mark Moyle, conductor of metal-mills, before the 1859 Committee.

Strictly speaking, the Sunday work in the Chatham mills is restricted to the rebuilding of the furnaces by gangs of bricklayers, who will generally be found at work

THE DOCKYARDS.   33

and who lets fall words of spiritual encouragement and admonition in his morning rounds among the beer-drinking sinners that are about him, administrative iniquity is sure to yield abundant fruit. The Captain-Superintendent in such a case believeth and hopeth all things of a professing Christian dockyard officer, and hopes and believes nothing of a non-professing Christian officer. Let the latter be ever so steady, indefatigable, and competent, the door of advancement is closed against him. Should he think of gaining promotion independently through the instrumentality of competitive examination, deterring hints are openly and boldly thrown out to him. If after this he perseveres, slander may be brought to bear, and then of course he succumbs. He is reported idle and wasteful on duty, and if in a leisure hour he has given his attention to some mechanical contrivance and taken the model home, purloining of the public stores is whispered. To crown all, his private life is spoken of. On the other hand, let a good-for-nothing intrusive officer be found holding forth on a Sunday evening to a crowd of dockyard workmen, and the pious Captain-Superintendent will not sleep before he has thought of some suitable step in rank to encourage one so worthy. He may remember that he has been applied to by the Secretary of the Admiralty for a deserving officer to proceed to Lyons to inspect the armour-plates contracted for by My Lords in one of their good-humoured moments, or to proceed to Liverpool, Glasgow, or Newcastle to watch the building of a ship of war for the Greeks or Turks. The street-preacher receives one or other of these appointments, although

---

during the hours of divine service, the cooling of the furnaces precluding them from doing the job on Saturday. On Sunday night the furnace fires are lighted, so as to be ready for the men on Monday morning.

he may never have seen either an armour-plate rolled or forged, or although, on going to the shipyard where the Greek or Turkish ship of war is building, he is obliged to confess in silence, perhaps a dozen times a day, his utter ignorance of right and wrong in shipbuilding. There is a story told of one of these incompetents, and it is worth repeating.

"How," asked the shipyard foreman, "do you want this done, Mr. Dockyard? There are three ways of doing it; one is this, the other that, and the third the contrary. Which is your choice?"

"The third," replied Mr. Dockyard.

"Very well," observed the shipyard foreman. "I shall jot down your determination, and read it to you, as I dare say you will not care to do it."

The memorandum was at once written, and Mr. Dockyard, nodding his assent, left the shipyard.

Some days after, Mr. Dockyard, after inquiring with some minuteness as to the other ways of doing the same thing, changed his mind in favour of one of the other ways, and ordered the alteration in the usual dockyard style, when the following short conversation passed between him and the shipyard foreman:—

The Foreman, fiercely—"No, sir, there shall be no change. The job was done by your order, and the Admiralty must now put up with it."

"Very well, sir," replied Mr. Dockyard as haughtily as a railway ticket-clerk who is applied to next day for the purpose of correcting an error in the change for a Brighton ticket; "I shall report your non-compliance with the contract in the usual manner."

A lengthy correspondence followed, the upshot of which was that the officer was mildly censured for inexperience, and recalled to his official duties in the dockyard.

THE DOCKYARDS.

Pious Master Shipwrights. Although the master shipwrights have little time for anything but reading Admiralty orders and instructions referring to the unconsolidated Admiralty statutes, and signing their names to official documents, it sometimes happens that where the superintendent is not morally of much account, the master shipwright is worth a great deal, at least in his own estimation. His piety as a rule is also of a narrow turn. At church on Sunday mornings he takes a note of the attenders subordinate to him, and in his walk across the meadow in the afternoon it is not to be supposed he will close his eyes to the abounding wickedness that prevails. Dockyard workmen observed in brawls, seen in beershops, or noticed with questionable associates of the other sex, all figure in the pious master shipwright's "black list." Such men being in spiritual darkness, and possibly also downright unbelievers, are not eligible for promotion at the master shipwright's hands. If hewers of wood and drawers of water, such they must continue to the end. Sometimes an Evangelical master shipwright will go beyond mere repression, and recommend for superannuation an individual who, if continued on the establishment another year or two—aye, sometimes another month or two—would be entitled to £5, £10, or £15, or more pounds annually during the remainder of his days. The men of known piety always get on smoothly, and if the truth is told, their work, whether manual or supervising, is never much inquired into. That a profession of godliness, proper as it is by men in all stations, should be made almost the sole dockyard ladder of preferment, must surely, in the opinion of all right-thinking men, be deemed the gravest of all the grave charges against the administrative dockyard system of the Admiralty.

The French Dockyards. Let us now briefly turn to the dockyards of other countries, that the objectionable nature of our system may be more apparent. The French dockyards are governed in a manner as near as can be the reverse of ours. The Minister of Marine is their autocrat. His pleasure is their law, and to his command all yield obedience. To the Emperor the Minister of Marine is accountable. If the Emperor resolves on any enterprise, the Minister of Marine is at his elbow, and on the instant, so well ordered are the French dockyards, that he can say positively—nay, can engage positively—to provide for this, that, and the other thing in a week, month, or year. Should the Emperor grumble, the Minister of Marine will reflect for a moment, so as to devise the means of performing in a week that which he thought of in a month. Yes, he can at all times meet the Emperor's views. The dockyard workmen, instead of being employed ten hours a day, shall work fifteen hours or even seventeen hours if necessary, and in addition, if the case is one of urgency, the factories and shipyards of the empire shall do all they can. On that understanding he leaves the presence of the Emperor and repairs to his office. Instantly the *Conseil d'Amirauté* is convened, the orders of the Minister are communicated, and it is the business of the *Conseil* to see them carried out. Whether a squadron, a flotilla, or fleet is wanted, the Emperor's plans will not be thwarted by those whose only duty is to serve him. The *Maritime Prefects* are instructed in plain terms, and those gentlemen know no such word as fail. Neither to birth nor seniority is their position owing, but solely to the confidence reposed in their ability and zeal. In peremptory terms the orders of the Prefects are in turn given, and the reluctant or disobedient officer or artificer would soon have to answer. There are no old gentlemen on the French maritime establishments who must go down to Brighton with the afternoon express, nor young gentlemen who must enter appearances in Rotten-row. The French dockyards in peace are on what may be called a semi-war footing—a footing which does not differ very much from that on which the allied army was after landing from the troop and transport ships in Turkey. Officials are not in the presence of an enemy, but they are constantly under less or greater expectation of being called on to prepare for one. They are, therefore, always thinking of their posts, and always ready to repair to them. None but those who have witnessed it or have been informed on competent authority can fully comprehend the spirit animating the French dockyard officers. They have no knapsacks, and therefore no Marshal's baton is in prospect; but all are conscious that alacrity, application, and skill will, some time or other, receive appreciation and reward.

<small>Our system in comparison.</small> How does our system compare with that of France? Lord Palmerston, let it be assumed, is apprehensive of war with some Power or other, and he sends for the Duke of Somerset. And what would be the sure result? Why, unless his Grace by good luck happened to have the Navy List in his pocket, he would not be a bit wiser than Lord Palmerston. With the Navy List in his hand, he would find out where this and that ship were at the moment serving, but he could offer no opinion as to what the seven dockyards could or could not do in a month or year, until, probably, after a week or even a month's consultation and correspondence with Admiral Robin-

son, Mr. E. J. Reed, and the master shipwrights. Our dockyards are always in the condition of a kitchen garden which the outgoing tenant has not cared to crop or weed. The Duke of Somerset knows something, Admiral Robinson knows something, and so does Mr. Reed and the master shipwrights, but the whole *personnel* of our dockyards is not up to the mark of one French Minister of Marine. Like the leading man of Welch and Margetson, or the leading man of any of the other great City warehousemen or merchants, the French Minister of Marine knows what everything is and where everything is. Ask him for cambric, shirting, sheeting, shirt-fronts, or No. 7 needles, and in an instant he will supply you, desire you to be seated, and talk about the weather. Then, as to the Whitehall and dockyard young gentlemen who part their hair in the middle being asked to stay after hours, the idea is too horrible! They could not do it, and they would not do it, and, what will strike most people as remarkable, the Admiralty cannot make them do it. In some odd way or other a dockyard appointment has come to be regarded in much the same light as the inscription of one's name for Consols in the books of the Bank of England. It is a provision. The public faith is solemnly and sincerely pledged to the young gentlemen who part their hair in the middle and look out for four o'clock daily in the public service. Why, then, should they be asked to stay after hours? What have they to gain by it? Why should they be pestered with an impudent attempt at menial degradation? If there is more than the regular amount of work to be done, why should not the Admiralty go into the market and hire the needful number of supernumeraries? Such is our system by the side of that of France; and it is easy to foresee that if ever our dockyard system has to bear the strain of a great war, the miserable sham will startle, if, happily, it does not shame us.

French and English Workmen. Of course it is not here recommended that the *personnel* of the English dockyards should be remodelled on the French plan. Such a thing is impossible. The British Admiralty will always have to deal with workmen who will not be driven but may be coaxed into working day and night without intermission until their strength gives out. Degraded as the dockyard labourers are, who, in return for their services, accept 13s. weekly and contingent superannuation, the compulsory labour to which the Frenchmen yield submissively would be resisted by noisy combination. The English dockyard labourers are still at bottom Englishmen, and, as such, must kick against coercion. Frenchmen, however, do not, and for one among many reasons, because they cannot. Were they to stand out and say half as much as the lowest class of Englishmen in like circumstances, deportation to Cayenne would assuredly be the lot of at least the more unruly. And so ordered are things in France that were the authorities to dismiss those who were not deported, the chances are they would be unable to procure employment in the same line again. The liberty of French workmen we in this country would deem oppression. In no sense are they their own masters, and in no sense are they competent judges as between themselves and their employers. To recommend, therefore, the remodelling of the English dockyards on the French plan would be absurd. But the French dockyards nevertheless present a striking and instructive contrast to our own. Nay, more, in the relative inefficiency of our dockyard system there exists danger. If the French Minister of Marine can in a moment use the great resources at his disposal as he wills, and every movement of the British Admiralty is clogged by red tape, indifference, idleness, and pretension on the part of the *employés*, the day may come when a state of things so unworthy of a free and all-powerful country will recoil upon it. In a word, the day may come when France may profit by our unreadiness.

<small>The American Dockyards.</small> The American dockyards are a mere adaptation of the French. Mr. Gideon Welles is to the American dockyards just what the Minister of Marine is to the French dockyards. Let President Lincoln determine on a new expedition, and Mr. Welles is closeted with him. When Mr. Welles was installed in his new office, his feeling must have been identical with that of any one installed in the responsible position of general manager of one of our great English railways. The department was made over to him, and if he is responsible to the President, it is in the sense that a railway manager is responsible to the directors and the shareholders: responsibility in both cases consisting in answering questions satisfactorily. Fully alive to this form of responsibility, it is much to Mr. Welles' credit—however much in other matters there may be said against him—that he applied himself to his new work as energetically as can be well imagined. With no pretension to any knowledge of dockyards, ships, or guns, he was an inquirer, without bias, open to such impressions as interested and disinterested people had ability to make on him, and he is now a notable example of the celerity and certainty with which, down to details, the naval system of a great country may be mastered and

THE DOCKYARDS. 41

directed by any clear-headed, plodding man. It is said of Mr. Welles by Englishmen who, in the course of the present unhappy war, have had abundant opportunities of judging, that the vast and intricate affairs of his department are at his finger ends. Just as a railway manager will tell on the instant what has been done, what is at the moment doing, and what is in contemplation along the several hundred miles of railway confided to him, so Mr. Welles is equally prepared. Where his ships are he knows, who commands each of them is linked inseparably to the name by some association or other that is inscrutable, and in one and all of the dockyards and shipyards, where ships of war are building or repairing, there is not a master or foreman that is not known to him familiarly. To-day he spends the forenoon in the bureau of yards and docks, and the afternoon in the bureau of construction and repairs. To-morrow his time is divided between the engineering, the ordnance, and the equipment bureaus. Next day is devoted to appointments with contractors for ships and other things, and the day after to important interviews with inventors. So his time passes from week to week and year to year. At first his task may have seemed as impracticable as the declensions to the young Latin scholar, or the French verbs to the student of maturer years, but now it is performed without an effort. Mr. Gideon Welles has the naval resources of the United States as well in hand as the French Minister of Marine has the naval resources of the French Empire. Is it not again, therefore, pitiful that of England what the well-known French committee said is still true, that "in England there is nothing organised."*

---

* *The National Almanack* (Philadelphia, 1863) gives the following official statement of the duties of the Secretary of the American Navy:—

Secretary's Office.—The Secretary of the Navy has charge of everything connected

What the American Secretary does not do.   Mr. Welles never rides in Rotten-row. Early risers have met him trudging to his post an hour or two before people usually are abroad, and he has been known to work on such occasions until after midnight. He has no set hour for receiving morning visitors, but is accessible at all times, even in his own house, to those with a good excuse for calling on him. Towards the middle of the day he is never wanted to help to make a House, and at four o'clock he is not looked for on the Treasury Bench in Congress, and expected to remain there until the motion has been made and carried

---

with the naval establishment, and the execution of laws relating thereto, under the general direction of the President. All instructions to commanders of squadrons and commanders of vessels; all orders of officers, both in the navy and marine corps; appointments of commissioned and warrant officers; orders for the enlistment and discharge of seamen,—emanate from the Secretary's office. All the duties of the different bureaus are performed under the authority of the Secretary, and their orders are considered as emanating from him. He has a general superintendence of the marine corps, and all the orders of the commandant of that corps should be approved by him.

The Bureau of Navy-yards and Docks has charge of all the navy-yards, docks, and wharves, buildings and machinery in navy-yards, and everything immediately connected with them. It is also charged with the management of the Naval Asylum.

The Bureau of Construction and Repair has charge of the building and repairs of all vessels of war and purchase of material.

The Bureau of Ordnance has charge of all ordnance and ordnance stores, the manufacture or purchase of cannon, guns, powder, shot, shells, &c., and the equipment of vessels of war, with everything connected therewith.

The Bureau of Medicine and Surgery manages everything relating to medicines and medical stores, treatment of sick and wounded, and management of hospitals.

The Bureau of Steam-Engineering, formerly attached to the Bureau of Construction, Equipment, and Repair, has been, in consequence of the great increase of the navy, made an independent bureau, and the Engineer-in-Chief made its head. The superintendence of the construction of all marine steam-engines for naval vessels, and the decision upon plans for their construction, belong to this bureau.

The Bureau of Equipment and Recruiting is another new bureau organised in consequence of the great addition made to the naval force. It has the charge of the recruiting stations for seamen, and of the furnishing them with the necessary equipments.

The Bureau of Navigation is a new bureau. The Naval Observatory and Hydrographical Office are under the charge of this bureau. It furnishes vessels with maps, charts, chronometers, &c., together with such books as are allowed to ships of war.

that this House do now adjourn. Mr. Welles does none of those things. When the President convenes a Cabinet Council the routine of Mr. Welles' duties is disturbed, but only to be resumed when the meeting is over. Nor at any time in the discharge of those duties is Mr. Welles reduced to the condition of the machine invented by Mr. Babbage. He calculates nothing, signs his name seldom, and writes little. Like our City merchants who buy and sell Consols and merchandise, Mr. Welles is a mere director: a man who says to this one, do this, and to another, do that. He controls; some commit his wishes to paper, and others see that his orders are carried out. Step to his desk and he shakes you warmly by the hand. He is perfectly at your service all the time you care to stay. But you have interrupted him nevertheless. That gentleman by his side is now busy extending the shorthand notes which have just been taken down. These notes embrace orders to Admiral Porter, instructions to Commodore Wilkes, and the acceptance of several contracts for the construction of ironclads. The moment the shorthand notes are written out, the clerks of the department begin their task, and those clerks are charged with the responsibility of transmitting the despatches to their respective destinations. Many London merchants conduct their business in the same manner, and at the year's end have no reason to be dissatisfied with the result.

American and English Workmen. Over the dockyard workmen the American Secretary has even less control than the British Admiralty over their workmen. In American labour, combinations extend to the dockyards, and the Secretary has no help for it but to treat the workmen as on an equality with himself, he being the buyer and they the sellers of labour. Should the workmen be unreasonable, either the work must not proceed or there must be an attempt to obtain workmen from other parts. On the contrary, should the Secretary be unreasonable, the workmen bundle up, and return to their employment when the Secretary has seen his folly. Strange to say, as will be shown presently, this free system—that is, workmen and employers opposing each other with all their might—answers admirably, and is infinitely the most productive to employer and employed. Its consideration will furnish one of the topics of a new chapter.

Chapter II.

DOCKYARD LABOUR.

The difficulty of organising Labour. The difficulty of organising labour, particularly in masses, is well known. It, in fact, constitutes the problem that great employers usually have to solve. A bad day's work, where there are many men engaged, may involve a loss of several hundred pounds; and a few weeks or months of bad work may ruin any enterprise or any well-to-do employer. The mere hiring of men is a task to which any one is equal, but employing them advantageously is an art in which few excel. Those who might be intrusted with the bringing together of thousands of skilled and unskilled workmen, those preparing estimates, drawings, and plans of work to be performed, and those versed in the mystery of raising money in large sums, would all, perhaps, be brought to grief in an attempt to make a profit out of the labour of masses of working men. The difficulty, of course, is in human nature, and success is to be sought in conciliating, enforcing, and occasionally in intimidating. Not long ago a well-known contractor, with 15,000 men at work, on the approach to his office of a deputation for higher pay, took a wager that he would balk them till next pay-day, which was another fortnight. His plan was the simple one of throwing down his hat on the office floor as the deputation were upon the threshold, jumping upon it, and swearing terribly. The leading man of the deputation looked at his companions and whispered that it was no use saying a word, "the governor being in such a passion." The wager was consequently won by the contractor. He could, up to a certain point, coerce his men—that great body of men—and the possession of the power was one of the chief elements of his success. He could also command a fair day's work, and in a high-handed manner dismiss skulkers, but at the same time he was always ready to yield on points of right and usage, and to respect the humblest hodman. Thus the organisation of labour to secure profitable and creditable results is an accomplishment of a high order. Its basis is a deep knowledge of human nature, and the superstructure the constant exercise of close supervision and common sense. Without these qualities there can neither be success nor satisfaction to employer nor employed, for, in the perfect organisation of labour in masses, the workmen have a great deal at stake.

The experience of Contractors. The experience of contractors in dealing with workmen of different countries places the difficulty of efficient labour organisation in a strong light. Germans use the spade and pick and wheelbarrow with great deliberation. Every spadeful is balanced on the wrist, and the back is straightened after every effort. The pick is raised slowly and awkwardly, and when barely erect allowed to fall with little or nothing more than its own weight. Last of all, the wheelbarrow is handled feebly, and pushed with a reluctant waddle. One day an English overseer, disgusted with the little progress the work was making, proposed contracts to the Germans on advantageous terms, in the hope of getting them to exert themselves in a proper manner. His proposal was accepted, and after measuring the job agreed on, he left them in the belief that the thought was a happy one to both. He was doomed to disappointment. Returning to the ground soon after, he was surprised to find no work in progress, and the men stretched on the grass below some trees. They had exceeded their usual earnings by a paltry sum, and preferred "repose" to further labour for the day. Unlimited earnings do not stimulate German navvies. Strange to say, Scotchmen, who are generally believed to stand in need of nothing but encouragement, are ill adapted also to contract-work. They are too painstaking and too thorough to succeed in that way. Scotchmen, as experience proves, distance all competitors in the character of their work, not excepting even Frenchmen, but they require incomparably more time than others, the Germans alone excepted. A Scotchman proceeds, not lazily nor awkwardly, but reflectingly, and the moment a doubt arises his snuff-box is interrogated, if he is a snuffer. If not a snuffer, it is a study to watch him in his perplexity. This deliberation, with untiring application, is what renders Scotch steam-engines, or the engines made by Scotch workmen, what Benson the watchmaker would call perfection in mechanism. In navvying, the infirmity to do well that which has to be done renders the Scotchman an inferior worker to the Irishman or the Englishman. He earns considerably less money than either. Than Englishmen there are none with a more perfect knack of bestowing just enough pains to answer. Place them on unlimited earnings, at fair remuneration, and where other men can do little more than starve they will be in receipt of an income in excess of all their proper wants. Americans are not behind, but on the contrary they are, if anything, in advance of Englishmen in adaptation and susceptibility to inducement when it is offered. Last of all, Frenchmen are found to excel only in little things. In mechanics they are greatly inferior, especially to the Scotch, the latter displaying infinitely more taste. In intelligence, earnestness, and quickness they are also inferior to Irishmen, Americans, and Englishmen.

The Dockyard labour problem. The dockyard labour problem is thus a grave one. And to add to its embarrassment there is in human nature a disposition to take advantage of special circumstances in all cases where exertion is required. The doctor for example thinks sluggishly, and not unfrequently erroneously, of a patient in easy circumstances, while he seldom fails to have constantly in his mind the struggling father of a young family whom sickness or accident has overtaken. Lawyers, again, have been known to allow easy clients to die intestate who never wished to do so, and few of them are proof against a fee, although the suit may be the most unnatural or oppressive that could be thought of. So with all classes, because it is a weakness that cannot possibly be overcome. When men, therefore, enter the service of the Crown in the dockyards, either as labourers, mechanics, or officers, it is natural for them to distinguish between the case of the Crown and the case of private firms. Were they entering the service of a private firm, that sense of justice which all men possess would incline them to act fairly by that firm, but, be it observed, no further than the personal standard of fairness each individual unconsciously forms in his mind. To this truth all unprejudiced and observant employers of labour must unqualifiedly assent. Although a workman can neither read nor write, he is dull indeed and useless if he cannot, by sheer instinct, judge, in labour to be performed, what is just to his employer and himself; justice to himself being measured with scrupulous exactitude by the treatment he receives and the wages he is paid. But to return. The disposition to act fairly by a private firm has one strong motive. Men employed by a private firm are anxious to remain employed, and to do so it is obvious to them that for the wages paid value in labour must be given. This is the great condition of the continuance of private firms, and it is one that the most illiterate workman perfectly understands. No doubt there are other motives than this eminently selfish one, but observation has long since proved them so weak and inoperative that they may as well be passed over. Men serving a private firm render a fair day's work for a fair day's wage chiefly because they do not want to shift about or be unemployed. On this truth considerably more light will still be thrown. Let us now turn to those entering the service of the Crown in the dockyards. What is the first inevitable reflection of such men, be their condition high or low, their minds cultivated or in the same state of nature as the glaciers of the Matterhorn? They are without that motive which alone binds private employers and employed together. The dockyards, such is their impression, are founded on a rock. No effort, that is no reproductive effort, of theirs is needed that the dockyards may be maintained. The dockyard superintendent is not in the position of a master hirer of their labour, be that labour skilled or unskilled, mental or physical, looking forward anxiously to the realisation of certain stages of the work in hand, so that his bank account may regularly be restored to a prudent balance. No such thing. If there is a deficiency in the dockyard accounts at the close of the financial year, be the cause of that deficiency what it may, Parliament, with a ready hand, will untie the national purse-strings and supply the needful. Sloth, therefore—or, let the plain truth be told, dishonest idling—that infirmity of our nature, wherever and whenever it can be safely practised, seriously and all but hopelessly complicates the dockyard labour question.

<small>Objections met. The case of Russia.</small> Before advancing another step, it will be well to meet one and all of the objections that may be urged by adducing a case where the evil is admitted and in course of being grappled with in a common-sense, refreshing manner. The allegation, let it be repeated, is, that men working for a Government in any form or capacity are bereft of the one motive to industry which renders private enterprise successful. Now for the substantial and unanswerable proof. It is provided by Russia. Russia, no doubt, wearied and disgusted with the wasteful and unsatisfactory outlays in its dockyards, has, to a certain extent at least, adopted the private labour system. An arrangement has been entered into with a Newcastle firm, substantially to the effect that Russia makes over to the Newcastle firm its Cronstadt dockyard, with workmen, plant, &c., for the construction of ironclads and the conversion of wooden ships to ironclads after our *Royal Oak* pattern. Russia wishes to possess its Cronstadt dockyard on the cheapest terms, and at the same time on the most efficient footing. To attain these ends it practically retires from the dockyard, and the dockyard, as regards labour, becomes a mere private shipyard, depending on the reproductive

character of its labour and its good management for future maintenance. Russia knows that if the experiment succeeds its ships will be built cheaper and more expeditiously, and that as a consequence an infinitely more useful dockyard will always be at its disposal. Then, success at Cronstadt would lead to the transfer of all the Russian dockyards into private hands and necessarily to an indefinite increase of naval power, contemporaneously, no doubt, with greatly diminished navy votes. Is not this an answer to all the objections that can be raised, whether Russia succeeds or fails? That Russia will succeed eventually and to the full extent desired no reasonable man will doubt, but the obstacles must for a time be great and unencouraging. For a time the Newcastle firm and the Russian Government may be as much mixed as the Elswick works and the Woolwich factory in the manufacture of Armstrong ordnance, but resolution and clear-headedness must in the end prevail. Out of the now hopeless Russian workmen the reciprocal dependence of employer and employed will in time be created, and the workmen taught to serve the Government as honestly and well as they serve private individuals.

Admiralty conciliation. Assuming the conditions of successful labour organisation to be conciliation, with force, and intimidation, together with proper supervision and the exercise of never-failing common sense, how far, let it be asked, are the Admiralty from the mark? What have the Admiralty done to conciliate the dockyard workmen? The Admiral, Commodore, or Captain-Superintendent of a dockyard directs and manages everything intrusted to him. Nominally his direction and management are regulated by the orders of the

Admiralty, but really he does as he likes, without being accountable to the Admiralty, Parliament, or the public. Practically, so far as the dockyard of which he is the superintendent is concerned, he is himself a Navy Board, or Board of Admiralty, making and unmaking those beneath him by his sovereign pleasure, and usually exacting an obsequiousness from those approaching him that would warm the heart of the King of Prussia, the last believer in the divine right of kings. In a word, his position is that of Admiralty commissioner and representative, with powers in full and proper form. He may interfere with and control the actions of every person within his jurisdiction, even to the details of artificer's work: he is, therefore, a master shipbuilder on a large scale, carrying on all the branches of shipbuilding without necessarily possessing a knowledge of any one of them. In nine cases in ten, an Admiral, Commodore, or Captain-Superintendent would be considered an inferior rigger, a useless sail or rope maker, a more than useless mast maker or rope maker, and a thoroughly incompetent painter, plumber, glazier, smith, carpenter, joiner, or engineer. Nevertheless he is a pretender in one and all of those branches, and it never has been claimed on his behalf that he is aware of the market or the dockyard price of what he commands. Of what is called business he knows nothing, and too often is a shameless bully. Too frequently Admiral, Commodore, and Captain-Superintendents are so strait-laced as to regard the dockyards committed to them as they would the quarter-deck when afloat, and if some of them had their will it is well known that the Articles of War would be read to the *employés* every morning beside a triangle and a cat-o'-nine-tails. Such is Admiralty conciliation. Such are the terms on which the labour market of the country is entered in competition with private firms. True, there are half-holidays after official visits, and on other great occasions, but they who fancy the working men of England—even the most degraded of them—are to be bought in that manner are mistaken. Gifts of that kind are received with much the same favour as exemption by the public from some obnoxious petty impost which Mr. Gladstone condescendingly announced but could not carry by resolution : or they are viewed in the light of schoolboy indulgences after a severe thrashing. A pretending bully in a cocked hat, who, in the solitude of the deck on foreign stations, has become steeped in the philosophy of the school of Zadkiel, brimful of quaint sayings, and oppressed with narrow-mindedness from not mingling like other people in the world, must, one would think, be a downright economic scarecrow. However eligible for fighting, or glibe of tongue or polished, no great employer of labour would set more value on him as the director and overseer of five or six thousand skilled and unskilled workmen than he would on an out-of-place flunkey. A great employer of labour, confiding in such a man, would hazard peace of mind and fortune. He would be setting over those strapping fellows who scale garden-walls in defiance of all protest, and after taking levels dig holes to find out what the subsoil is, a man so much beneath them in a professional sense that they would at once relinquish their positions. Even the Irish hodman would take offence at such a man, and embrace an early opportunity of accidentally dropping bricks or lime upon him. Conciliation forsooth! Let it with all respect be said, if there is any class that in unfitness for the ordinary concerns of life exceeds all others, that class is naval officers. As a class, on land they are, with few exceptions, useless and intolerable, however estimable and useful they may be, and really are, at sea.

Admiralty force and intimidation. If in the dockyards no trace of conciliation can be found, there are certainly no evidences of force or intimidation. One may walk a day or a month in the dockyards and at every turn observe wrong-doing or something out of place, but neither force nor intimidation will be witnessed. If the Admiral, Commodore, or Captain-Superintendent stops a labourer and tells him on which shoulder he ought to carry a plank, or a blacksmith in the smithy and tells him how nails are made in China and Japan, " as he himself has seen," the smith and labourer seem ready to fall upon their knees before the supreme being that has deigned to notice them. And so with all classes of dockyard workmen, and all grades of dockyard officers. Under such circumstances force and intimidation are impossible, because where all seem willing, where all seem broken, where all appear more servile than the Russian serfs or the Carolina negroes, there cannot be, nor ever is, a fault to find. What may have been seen amiss admits not of one but of several oily explanations. Therefore, neither force nor intimidation exist in the dockyards. Those great resources of all employers of labour have not yet taken root in the dockyards, because, happen what may, in the dockyards no one feels aggrieved. If the pitch-pot boils over and a stack of timber is consumed, whose timber is it, or whose pitch is it? If to light the pitch-pot fire some valuable shipwright's moulds have been used, whose loss is it? The man who so lighted the fire, who allowed the pitch to boil over, and destroyed the stack of timber, would tell his mate that the loss is good for trade. So his mate would think, and the Admiral, Commodore, or Captain-Superintendent, should he write to Whitehall on the subject, would inevitably, by way of tag to the statement of the accidental loss, bring the heroic exertions of some fellow or other to their Lordships' notice. In the establishment of a great employer of labour accidents of course will happen, but accidents there are accidents. Carelessness or remissness are both punishable in one of two ways: either the pay of the offender is stopped, or he is turned, if not kicked, outside the gate. Let a workman be wasteful, and either a foreman's or an employer's fist will be in his face instantly. Those who have attended the payment of crowds of navvies will have known cases where an employer with a spite at some fellows has attended in the hope of their committing themselves by noise or otherwise, that he might have an excuse to turn to and thrash them. They who employ labour must either for the time lower themselves to the level of the labourers, or they must elevate the labourers to their own level, if they wish things to move sweetly and profitably. A cocked-hatted naval officer is too conceited and incapable to try the experiment either way.

<small>Admiralty supervision and common sense.</small> Of course there is no such thing as Admiralty supervision, and no exercise of common sense, or the Admiral, Commodore, or Captain-Superintendents would have been drummed out of the dockyards long ago, with their authority restricted to the flag-ship or the ships in commission and ordinary. For such men the dockyards are an unsuitable and improper place. They are incapable of counselling or recommending in dockyard matters. If we want to know something of mixing mortar, let us get into conversation with the first bricklayer. If we want to know something of boating or of mackerel, let us ask a light from a Brighton or Margate boatman. So, if we want to put five or six thousand men to work at Chatham, as many at

Portsmouth, and as many at Keyham, let us rationally confide in people who know something of five or six thousand workmen, what they can and what they cannot do, and what they might and what they must do. Then supervision may be dispensed with, although its exercise would still be useful. Then the Admiralty may or may not possess the brains and ordinary intelligence of common mortals, although the possession would by no means be without its value. That at present dockyard labour is without a rudder or a helmsman is another to the many existing proofs that all the money that may be lavished on the dockyards by an unthinking and really reckless Parliament will neither secure results commensurate with the outlay nor with the naval requirements of the country.

The existing labour system. The conditions of successful labour organisation not existing in the dockyards, and the dockyards being consequently conducted on principles at variance with propriety and common sense, it will be well to fortify this severe judgment by a brief inquiry into the dockyard labour system as it really is. Shameful, positively shameful, is the history of the dockyard labour system. During the long war in which this country was engaged, the dockyards were the mainstay of the Crown and successive Governments. Within them hard-up peers' younger sons and useful commoners browsed and fattened, and in return for wealth often more than questionably acquired, they vigorously sustained the public policy. The committee of 1804 exposed many of the abuses that then existed, but, as in all other instances of dockyard inquiry, they were precluded from speaking of acts which Parliament would have been bound to notice and the courts of law to hear.

Honest dockyard witnesses, on all occasions of inquiry by commission or committee, when closely pressed have either spoken freely, on the understanding that their statements should not be reported, or they have felt disposed to do so, but have not had the hardihood to name their terms. One well-authenticated case will suffice, which came before one of the recent committees. A dockyard master shipwright was being examined as to the operation of what is known as the task and job system, and in answer to the direct and perhaps unexpected question whether cases of downright robbery had not frequently come before his notice and been passed over, answered that if the committee would cease writing, and make no use of what he said to his prejudice, or to the prejudice of any one, he would mention several such cases. His terms were acceded to, and one of the cases related to a large quantity of deck-planks on which several gangs of shipwrights had been at work for a week. These deck-planks, on being finished, were stacked against a shed to be convenient for the measurer when he made his round, and they were duly measured and the shipwrights paid. The job not being finished, the shipwrights continued working, and next week the planks which had been measured and paid before were turned round so as to hide the measurer's chalk-mark, and re-measured and re-paid. All the shipwrights employed on the deck-planks were of course aware of what was done, and so was the master shipwright, who testified the fact before the Committee; but conspiracy and robbery are dockyard usages as old as the privileges of the City Livery, and they are therefore winked at and forgotten. The other cases mentioned by the master shipwright were of the same character, and it is needless to repeat them. Even more barefaced things were done during the long war than getting deck-planks twice measured the other day, and one of the most profitable conspiracies that then existed still to some extent survives—namely, receiving and passing improper contract stores.

Task and job work in the Dockyards.* Task and job, as the words imply, is a method of tasks and jobs for certain sums. The Committee of 1804 based their objections to the system chiefly on the fact that all the superior dockyard officers were more or less directly or indirectly interested in the earnings of the workmen. That is to say, they were more or less pecuniarily benefited, just as the leading dockyard men are still. The leading dockyard men at the present time share the earnings of their gang, and yet they are the supervisors of their gang, its overseers and check as between the workman, the Admiralty, and the public. A position more trying, morally, is incon-

---

* We are of opinion that the system of task and job, as at present carried on in the dockyards, is open to great abuse. It is based upon schemes of prices consisting of 94,762 items. This can hardly fail to lead to numerous errors in priceing the work done. Another grave objection is the multiplicity of measurements required.

If task and job is to be continued, we would suggest that it should be employed only when ships can be divided into sections and the exact amount of labour to be paid for in each section estimated in the constructor's office.

We recommend that with the submission for building a ship the Controller-General should also submit a carefully prepared estimate of the entire cost of the ship, distinguishing materials, labour, fitting, rigging, and engines. That in all cases where alteration or repair of a ship is proposed a similar estimate should be prepared.—Report of the Commissioners; 1861.

The committee think that the facts which they have adduced respecting the result of the system of task and job afford sufficient evidence of that system having hitherto been disadvantageous to the public service in the manner in which it has been carried out. They are fully alive to the value of a system of task and job when carried out as usually done in the private trade; and it has evidently been the intention of the Board of Admiralty, at the various times when they have had the system under their notice, to give orders for its use in the yards on the same principle; but from the want of experience and knowledge on the part of the professional officers of how that system was worked in private trade, they have failed to secure its advantages to the Government.—Page 22, Report of the 1859 Committee.

ceivable, and it is not surprising that task and job was condemned by the 1804 Committee. The Admiralty of the time were, however, opposed to innovation, and two years later—namely, in 1806—the task and job question was remitted to the Commissioners on the Civil Affairs of the Navy. These gentlemen qualified the censure of the 1804 Committee by recommending task and job for new work, and as far as practicable for repairs. Accepting this qualification, a committee of master shipwrights was appointed, and a scheme of task and job prices agreed to went into operation in 1811, and with modifications and additions continued in operation until 1833. This committee of master shipwrights, it may be well to state, overlooked one important recommendation of the Commissioners of 1806, which to this time has not been rectified by adoption by any Board of Admiralty—namely, that in all schemes of prices explicit reference shall be made to the completion of ships of war for certain sums. The abuses under this omission frequently attracted the attention of the public, and in 1833 task and job work was superseded by day work and day wages. In 1854 day work and day wages was superseded by task and job work again, which now continues.

Day work and day pay in the Dockyards.*
Day pay and day work did not answer any better than task and job work in the

---

* If the sectional system cannot be applied in the dockyards, we recommend a return to day pay with increased superintendence, and that task and job should only be resorted to in cases of great emergency, and even then should be confined to the building of ships, and should not be applied in cases of fitting, altering, or repairing.—Report of the Commissioners; 1861.

The same facilities for deceiving the measurers exist under day work and day wages as under task and job; and when applied partially or to a portion only of the work executed, these facilities are increased by the feeling existing against it. The committee consider the objections to the system of day pay under check measurement so great that it would be advantageous to the service to discontinue it, and supersede it by another system founded on sounder principles—namely, on the system of the private yards.—Page 24, Report of the 1859 Committee.

dockyards. If under the task and job system the public money was stolen openly without a single thief being censured or gracing the assize or Old Bailey, the system of day pay and day work effected no saving. The men employed on day pay and day work either could not work or would not work, and officials casting about for reasons to account for the extraordinary state of things that had become notorious arrived at the rather odd conclusion that the supervision was excessive, and that with less supervision and a system of check measurements, more satisfactory results might be anticipated. Accordingly, in 1847, the supervision staff was materially reduced, and each man's work was measured daily. Two years later—namely, in January, 1849—the Board of Revision recommended a still further diminution of the supervision staff, and an increase of the gangs to twenty men and apprentices. Experience has since proved that neither check measurements as a substitute for excessive supervision, nor supervision with or without check measurements as a means of obtaining a fair day's work, or as a means of constructing or repairing ships of war at moderate prices, is in the least satisfactory.

What these failures show. Now, what do these failures show? Do they prove that the dockyard labour system is a model system that all great and small employers of labour will do well to copy? Quite the contrary. They demonstrate the fact that all the dockyard labour schemes have from first to last—that is, up to the present hour—failed most miserably. At this moment the task and job system now in operation opens as many avenues to fraud as existed at the beginning of the present century. The Committee of 1854 discovered that at task and job some men could easily earn 24s. or 25s. a day, while others, equally good workmen, might sweat from morning to night at task and job and earn with difficulty 3s. Of course, the most obsequious but really the most worthless fellow in the dockyard,—the fellow, probably, who is making improper overtures to the Admiral, Commodore, or Captain-Superintendent's cook, —gets the 24s. or 25s. work, and the honest man, who fights the hard up-hill battle of life with a numerous family, gets the 3s. work. Then the pious leading man of a gang may still make all the money for himself and men that his conscience will allow. Who for a moment would doubt the integrity of the fellow who holds forth in a barn on Sunday morning, and at the dockyard gate on Sunday evening? True, he does not work himself and only receives his share of the earnings of his gang, the amount of those earnings being, however, determined by himself, but still he is without reproach, a man in all respects worthy of the next frontispiece and memoir in the *British Workman*. Then, last of all, the men charged with the duty of receiving and approving contract stores still enjoy all the old-fashioned opportunities of being corrupt. They may pass timber at complete variance with the specification, and which is of no use in the dockyards, because wanting in sufficient length, breadth, or thickness, or because it is full of sap or rotten. They may receive tools which instead of being all steel are steel only at the points. In a word, there is still no check on them. For all the public know, like some of the American receivers of shoddy articles, they may be in league with those who supply the shoddy. Thus the dockyard labour system is an abuse as flagrant and venal as the rotten borough system that the Reform Bill swept away. It is a system worthy of the Turkish Pachas or the Chinese Mandarins, but unworthy of free and enlightened England. By all means when Parliament assembles let an attempt be made to obtain the dockyard "Black Lists," for these are only needed to brand the dockyards with eternal infamy.

Limited earnings.* But these strictures on dockyard labour will no doubt be met by the statement that "My Lords Commissioners of the Admiralty have been enabled, by the consideration they have given to the subject, to correct or lessen the abuses of task and job work, by the very satisfactory expedient of limited earnings." The statement would be untrue. In the first place, the practice of unlimited earnings is not thoroughly carried out; and in the second place, the practice can never be satisfactory. What is implied in limited earnings as pretty generally carried out in the dockyards at the present time? Take the case of the leading man of a gang. This man can only point out to the measurer, on behalf of his gang, a certain amount of work for which payment may be claimed. That is to say, if the fellow is a swindler he can only swindle up to a certain point. Turning to a stack of deck-plank, he might say, "These have been all dressed by the gang, and the money to be paid to the gang exactly tallies with the maximum of the limited earnings they are entitled to receive." "Very well," replies the measurer, and he certifies for the amount. But were these deck-planks all dressed this week? This is the weak point. May not a portion of this work in reality have been

---

* The men, though still paid under the scheme of prices for task and job for the work done by them, were put upon limited earnings per day, and to carry out this the hours of labour per week were lessened; but if the men executed more work, so that their earnings exceeded the limited amount, they were not paid for the excess of work.—Page 20, Report of the 1859 Committee.

DOCKYARD LABOUR. 63

paid before? Let us next pass to the question that the practice of limited earnings cannot be satisfactory. This man, let us say, is an engineer, and his limited earnings, let us add, are 7s. a day. He can earn 7s. a day, but not a farthing more. Still he is on task and job, and it may happen that to-day he will exceed the proper limit and to-morrow fall short of it. How does that affect him? In this way: that on the day of the excess he is not paid for the excess, while on the day of the deficiency he receives no more than what he has really earned. Can such an unfair system be either satisfactory to the Admiralty or the dockyard operatives? Impossible. How is an operative to be affected to the Government he serves, if when he, by task and job, earns 10s., he is paid no more than 7s. because that is the "limited" maximum his class attains, while if he, by task and job, earns 5s., he is only paid 5s. because he has not attained the "limited" maximum? The butter is all on the one side of this arrangement, but still it is said to be satisfactory. Then again, here is a really good mechanic, a man who anywhere is worth his 15s. daily without the least apparent effort on his part. How is he affected by the limited earnings? Let us say the maximum of his class is 7s. Going into the dockyard in the morning and exerting himself, this man's seven shillings-worth would be finished by twelve o'clock. Is he in such a case at liberty to leave and return to work next morning to do the same thing? No. To do so would be a breach of the Admiralty orders. One of two courses is always before such a man: either he may give half a day's work to the Government for nothing, or he will make the work that he might finish by twelve o'clock last until the usually knocking-off hour at night. Of course such a man would not work for nothing, and he spreads his half-day's work over the entire day, and his idling about, gossiping, and sitting down is not only a bad example to the men about him— who are on day work and day pay, and there are still many such—but brings discredit on the dockyards. Still, it will be said, the limited earnings system works satisfactorily. It does no such thing. It is as much a failure as any other system that has been tried or that may be tried if ingenuity can possibly invent further means of countervailing the incompetency of a bad master. Than the system of limited earnings, none is to be thought of so well calculated to discourage and eventually spoil both good and bad workmen. And if it is the case that in human nature there is a proneness to do to others as they do to us, it is no marvel that a Government mean enough to deprive workmen of their excess earnings, these often being accidental, should meet, and that not unfrequently, with workmen mean enough to steal brass and copper in large quantities.

Additional abuses. But there are still additional abuses incident to the limited earnings system, and, in fact, to all dockyard labour systems. One of the more prominent of these is the system of equal pay. Here is a gang of twenty men who work together under a leading man who shares their pay. Shares their pay! Yes; these twenty men are all paid alike, although some of them are striplings and some of them are worn-out fellows verging on the threescore years and ten. One-third of the whole number do twice the work of the other two-thirds, and yet all share and share alike. This is practical socialism with a vengeance, and in a Royal dockyard too. This is what Robert Owen would have liked to witness, but it is what no good workman relishes. Want of labourers is another shortcoming. This gang of twenty men we shall say are shipwrights, and under the limited earnings' system it is found to answer to oblige the gang to perform their own labourers' work, because it keeps the men as well employed as possible without putting into their pockets any money. They are to plank the sides of the *Duke of Somerset*, or to remove the store-rooms from the *Admiral Robinson*. To do the first, poles and planks have to be brought and transformed into stages; and to do the second, no end of dirt and lumber must be cleared away. To the degradation of doing these things the shipwrights must submit. Their skilled labour must be put to these purposes, although labourers might be hired, seamen of the reserve brought from the hulks, or convicts from St. Mary's Island, Chatham, or Portsmouth prison. No wonder, as the 1859 Committee reported, page 87, paragraph 949—"The Committee consider that the present arrangements have failed to secure IN ALL CASES the entry of the best men that could be obtained." No wonder that the best men, who can get a living elsewhere, shun the dockyards. No wonder that the private establishments should attract the engineering and shipbuilding talent of the country, and leave to the dockyards the men who can submit to the degradation of labourers' work—the men who will accept the practical socialism of equal pay and the practical fraud sanctioned in the loss of sums honestly earned in excess of the standard limit. What, however, is wonderful, is that men, and those men speaking officially, should have the effrontery to affirm that the workmen and the workmanship of the dockyards are superior to those of the private firms. The statement is untrue. As well at once assert that freedom and dignity are inferior to restraint and degradation.

Superannuation.* Still, as even Sodom and Gomorrah had their good people, so in the dockyards may be found a few skilful, upright, hard-working men. Why any such are to be found in the dockyards is to be accounted for solely by the pleasing prospect of superannuation with half, three-quarters, or full pay, according to the length of service. These men are of that quiet, unambitious turn we sometimes meet with, and were we to seek to know their character, any unkind feeling we may before have harboured to the system of providing for the old age of workmen rather than requiring them to provide for themselves is roughly shaken. These men give themselves soul and body—the soul must be included—to the public service, deeming their devotion a fair equivalent for present and prospective certainty and benefit. Their like are to be found among a pretty numerous class of public creditors, who work and hoard until their purchases in the funds yield a clear £1, £1 10s., or £2 weekly, on which they manage to subsist afterwards, enjoying the river-side and rod, the sea-side and the line, the fields and nature, the mechanics' institute, books and newspapers, or the fireside and the religious training of a family. Know such men, walk into their cottages, observe the contentment and comfort that prevail, and it

---

* 940. The committee have given much attention to the subject of superannuation to the workmen of the dockyards, and they are very strongly of opinion that the present system has an injurious effect upon the service, tending to protect an inefficient man, and prevent him from being discharged, because the supervising officer considers that after he has served a few years he has a claim upon the service, and that he would lose by his discharge.

941. The wages paid to the dockyard men are considered to be lower than those paid in private trade, and partly on account of their future superannuation.

942. A feeling therefore exists that they are deprived of something which they have earned, and which is their due, if they are discharged: and this leads to an unwillingness on the part of the officer to recommend them to be discharged, or even to report them to their superiors for misconduct, lest this report should lead to that result.—Report of the 1859 Committee.

must be confessed that to such, superannuation, like mercy, blesses both the giver and the taker. But pass from the cottages of these worthy workmen to the cottages of the drunkard, or of the man who, when the chance presents itself, carries off tools, or any other thing of value. Too often he is under the dominion of a forbidding scold, who, in addition to her other bad qualities, is slovenly and unthrifty. That man who has just stepped out, a curse as great as strong drink to the British working classes, is the tally-man of the street. He holds a perpetual mortgage over the scanty earnings of the family. Beginning with the sale of a shilling cotton handkerchief, worth threepence, to be paid in twelve equal weekly payments, he is now the sole clothier and haberdasher, and at a moderate computation he robs the dockyard workman to the extent of 10 or 15 per cent. of his earnings. There, however, the evil does not stop. This fellow, a regular visitor in the absence of the husband, is immoral. He is a fiend in human shape, and his steps are marked by the moral ruin, the conscious degradation, and the destruction of the peace of families. This fellow throngs the brothel, the workhouse, and the prison with his victims. But to return. However well bestowed superannuation may be in the case of the workmen named before, is it bestowed worthily in such cases? Why should improper persons be encouraged in wrong courses by the reflection that Government will provide for them in their age? Why should dockyard workmen, like other workmen, not be required to form habits of providence and self-dependence? Why, in a word, should Government offer the bounty of superannuation, to its own and the public loss, and to the positive detriment of those whom it designed to serve?

Originally a bribe. Superannuation was originally a complement to appointments by favouritism, or for corrupt reasons, and it has been said that at one time it was intended to carry the abuse still further, by allowing pay to descend to heirs male lawfully begotten during one or more generations, as was, and still is, the custom in creating peerages for distinguished services in those cases where the noble fledgling is too poor to support his title in a proper manner. But, on reflection, it must have seemed unfair to rank to think of such a thing, and accordingly the generosity of the Crown and Government was restrained. Beyond providing that the happy nominees should be cared for during the full term of their natural lives, there was nothing done. Considerate punctiliousness! But for it the dockyard poor law would have been as formidable a social evil as the ordinary means of ministering to the unfortunate of all classes. Surely with the knowledge of this deliverance, rank will hereafter be held in as high estimation as the office of Garter King of Arms, the incumbent of that office having disposed of the recent knotty questions of Court precedence. But even in the modified form in which superannuation has descended to us it is a monstrous evil. This man is master shipwright. He entered the dockyards twenty or forty years ago, and that is all that is known of him. Whether he was pitchforked into the service or rose meritoriously is now a matter of indifference; the smiles and favours of such a man being all that dockyard people at present care for. He occupies a handsome official residence, pays no rent, burns the public coal and gas, is neither called on for poor-rates nor taxes, and let us say his salary is £600. For all this he occupies his office during official hours, and signs his name. That is his duty, practically his sole duty, although he is master shipwright. Surely this man is greatly overpaid. But when he retires from the service he will receive £600, £500, £400, or £300 annually during the remainder of his days. This will be his superannuation. This will be his extra pull upon the public. This other man is a shipwright, not a master shipwright. The earnings of his gang average 7s. or 8s. daily. Whether young or old, competent or useless, this man receives this average sum, and when he is worn out he retires on £50, £75, or £100 annually. This is his superannuation. This is his extra pull upon the public. If you express your astonishment to him he will be ready with his answer. He will tell you that while his average earnings were 7s. or 8s. daily, shipwrights in the private yards were receiving 9s., 10s., and it may be 11s. Therefore, he will argue, the superannuation in his case means no more than the deferred payment of just claims. He, however, knows that he is only telling half the truth. True, his earnings are less than the shipwrights' in the private yards, but the latter may be three months idle throughout the year. On rainy days and snowy days the shipwrights in the private yards do not work, and no work no pay is the motto of the private firms. In the dockyards, when it rains or snows, the shipwrights are sent into the sheds or on board the ships, and by that means never lose a day. Nor do the advantages of the dockyard shipwrights stop there, for when sick they are attended by the dockyard doctor without charge, and supplied gratuitously with all he orders. They have, besides, a sick allowance almost, if not altogether, equal to their average earnings in many cases. To the shipwrights of the private yards there is, of course, no sick allowance, and neither medicine, comforts, nor medical attendance.

So, after all, the dockyard shipwright may be said to be better off than the shipwright in the private yard, and yet he retires on full, three-quarters, or half pay. For obsequiousness, limited earnings, and other things, he receives superannuation, while his services, such as they are, command much the same equivalent in constant work, medical attendance and comforts, and sick allowance, as the ordinary market rate of shipwright labour.

Its gross injustice.   Of the gross injustice of superannuation it is not difficult to judge, apart from its encouraging improvidence among the great majority of the dockyard people, and from its being an indefensible and vulgar bribe. Let us view it from a taxation and an employer-and-employed point of view. Men enter the service of the Crown in the dockyards, and, as has been shown, they are practically as well off as other people in the same private stations. They should consequently acquit themselves and stand on precisely the same footing as others. But they do not. Once in the dockyards they are recognised as fixtures, in exactly the same sense as one's wife is a fixture in one's household. Dockyard hands are entered for better and for worse. Whether they shall do well or ill they alone must decide. The Crown, by its agents, puts only the most paternal pressure on them, lest they should come to harm—that is to say, be turned off, and forfeit the superannuation. The Crown therefore spoils them. Now, who are these pampered hirelings? What do they render to society that society could not provide itself with? The answer must be—Nothing. What is urged in defence of retiring allowances to the officers of the army and the navy who in the service of their country have braved the perils of unhealthy climates or risked their lives in battle, cannot reasonably be heard in support of the dockyard people. The Crown might as well superannuate the London coal-heavers, and with much more propriety all the professional men who have devoted manhood, with inadequate return, to the doing of those things which otherwise would have remained undone. Then is not the toleration of the idleness, incompetency, and infirmity of the dockyard people, lest they should lose their superannuation, or lest the cup of superannuation should not be filled to overflowing, as monstrous as not laying the birch across a naughty boy's back in case it should interfere with the enjoyment of his dinner, or as monstrous as refraining from giving a drunken policeman or cabman into custody for fear the one would be suspended and the other get his licence cancelled? In fine, the employer who would treat his workpeople as the Crown does the dockyard people would very soon either have his creditors about him, or his next of kin inquiring by the usual commission into his state of mind. Then, regarded from a taxation point of view, there can be nothing conceivably more monstrous than the superannuation of the dockyard people. It is a downright robbing of Peter to pay Paul. To pay the superannuation of the dockyard people the working poor of England are taxed in many of their daily necessaries, and yet the dockyard people are the better off, and incomparably the least useful of the two. Why the Crown has never thought of ordering Union uniform for the dockyard people, that their true condition in life of paupers on the public rates should be known, is an omission that cannot be remedied a day too soon.

Admiralty disposition to effect a change. Of late the Admiralty have manifested a disposition to abate this monstrous evil. Like President Lincoln, they have been putting down their feet. Men are entered more freely as hired men and less freely as drones in the national hive. These hired men, of course, occupy the same position as all others of their class, and may be paid off when idle or incompetent or when there is nothing for them to do. But the Admiralty itself is an effete organisation, and therefore incapable of dealing resolutely with this vital question. The whippers-in of the party are still at liberty to listen to the representations of the members of the dockyard towns, and the great politicians of the dockyard towns pull the wires that wring compliance from the Secretary. As the case at present stands, for one hired man employed, one and a half if not two men are received on the establishment, so that the getting rid of the drones is still practically adjourned till doomsday. One of the earliest and most strenuously supported resolutions in the House, when Parliament assembles, should be framed to meet the evil. Lord Palmerston's nonsense about Parliamentary interference with the prerogatives of the Executive should be peremptorily silenced, and the entry of another individual on the establishment of the dockyards, be his condition high or low, formally discouraged, unless on the conditions of good behaviour, competency, and payment in full by salary.

English shipyard labour.
English shipyard labour is in the main uniformly organised, but in a sense it is still in a transition state. That is to say, while in the great establishments the division and efficiency of labour is a constant study, many of the small establishments have not yet succeeded in emancipating themselves from the dockyard practice, which only a few years ago was deemed the model of propriety,

not in England merely, but in Europe. France and Russia and the minor naval Powers were then not less slavish copyists of the English dockyard system than the English shipyards; and very oddly, it was customary —a custom not altogether yet exploded—for England to justify some of its dockyard usages by reference to some of those of France; for France to justify some of its dockyard usages by reference to some of those of England; both losing sight of the fact that the French system was a mere reproduction of the English in those particulars. By and by it will be seen, when the question of naval power comes to be considered, that there have been more delusions of this kind. France has constructed fleets because England has done so, and the course of France has led England to repeat the proceeding; the armaments of both countries moving in a vicious, and stupid, circle, without the least reference to the possibility of manning or using the constructed ships.

<small>The proximate cause of change.*</small>   The change from dependence on the English dockyards, both by the ship-

---

* I will first speak of the two articles, wood and iron. They represent two distinct orders of ideas in ship construction, as well as in political economy. The acorn that we plant to-day will be squared into logs next century. If we cannot wait the maturity of growing acorns at home, we must bring timber from abroad. The expense is enormous, and out of all proportion to the price of the timber at a distant shipping port. Then the cargo may not turn out well, the best timber may not be of the desired length, breadth, or thickness, and the worst, full of sap and useless, may in other respects be of the length, breadth, and thickness needed. So timber is a commodity of uncertain quality, and, as a rule, the longer it is kept the more uncertain its quality becomes. Will it be believed that the other day, notwithstanding the enormous quantity of timber in Deptford Dockyard, enough of the right sort could not be found to complete the stern fillings of the *Favourite* corvette? This uncertainty in respect to quality not only enhances the price of good sorts, but creates a disposition even in the dockyards to make shift with what is at hand. On no other supposition is the hopelessly rotten state of the ten-year-old frigate now in Keyham Dockyard to

yards of England and the dockyards of Europe, may be said to be wholly owing to the use of iron. As long as ships were built of wood, people flattered themselves that

---

be accounted for. But, here is an immense raft of logs floating down the St. Lawrence to Quebec. It is the hope and fortune of an entire backwoods settlement on the Ottawa, or of a colony in the French country. It is the result of a winter of exposure and hard labour. The raft is stowed on board ship, and landed at Woolwich or Chatham. It is again the hope and fortune of a settlement or colony. It is the source of smiles and comfort to hundreds if not to thousands. From the timber to the finished ship of war there is wide interval; so wide, indeed, that some ships have been seven years and more in construction in the dockyards. The logs are unmanageable, the cutting of them up tedious, the shaping of the parts laborious, the jointing of them slow and unassuring, and the eventual securing and binding of the whole of the million fragments in the finished ship of war by no means puts an end to the vocation of the joiner, the shipwright, and the caulker. On the contrary, the construction of wooden ships of war is an endless source of labour. A wooden ship of war is never finished, and never can be finished. While the shipwright makes good one part, decay is hastened in another by the want of air and other causes, so that renewal ought to move always in a constant circle. The advocates of wood in ship construction are therefore the advocates of waste. Timber in ship construction is to them a means of employing labour. It is a means besides of continuing intact the dockyard system. On the one hand they set at nought a well-ascertained and enlightened general principle, and on the other hand they oblige the country to cherish and continue the possession of that which advancement teaches us is neither worth cherishing nor possessing. Wood is the most troublesome, unfit, and costly material out of which ships of war can be formed; and it is nonsense to suppose that the dockyard towns are to be supported by a disguised system of outdoor relief.

Iron does not spring from acorns, nor yield to the backwoodman's axe. To possess it we have to dig at home, and neither buy nor carry from abroad. Staffordshire, Lanarkshire, and other counties of the kingdom are iron counties. Their ore, furnaces, forges, and mills invite us, and constitute one of the great tributary streams of our national wealth. The labour required to chop the pine, clear away the brush, square the log, drag it on the snow through the Canadian forest to some watercourse, assist and watch its progress over rapids, and guide it safely in raging inland storms until a shipping port is reached, seems almost ten times greater than that involved in bringing iron ore into a working state. So to make use of iron in place of wood in ship of war construction, is to prefer that which costs us least. It is to adopt and apply a principle (to which Englishmen owe so much) in a new direction and with the same good results. The less that labour separates effort from result, the greater the share of comforts that falls to the mass of rich and poor. Iron, besides, is indestructible or nearly so; and it may be fashioned readily, even in ponderous masses, into any shape. The time involved in bringing timber from the backwoods of Canada to Woolwich or Chatham would suffice for the complete construction of a formidable iron ship of war in such an establishment as that of the Thames Iron Shipbuilding Company. As fast

shipbuilding science had been carried to its furthest limits, and better and worse classes of merchant ships meant little or nothing more than near or remote approaches to the dockyard way of doing things. A ship of more than usually heavy scantling, and with a variety of foreign timber judiciously distributed in all its parts, might have fairly claimed to be frigate-built, and was sure to be toasted as such at the launching banquet; while a ship of unusually light scantling, and in which plain fir and elm, with a sprinkling of American white oak, only figured, was far below the dockyard standard, and comparatively little thought of. When the history of British shipbuilding comes to be written as it deserves, and no doubt some day will be, the veneration for the dockyards and the pertinacity with which the dockyard fancies for timber, fastenings, and other things were supported by the shipowners and shipbuilders, are sure to be amusing. It is not, perhaps, too much to say that the devotion of the shipowners and shipbuilders will afterwards find a place in the well-known history of popular mistakes. Instead of doubting and inquiring for themselves, they accepted on trust all that was done in the dockyards, and all that emanated from them.

---

as furnaces, not seven times but seventy times heated, can soften the bloom of iron until the drops fall from it like water as it is drawn along the floor, may angle and other iron be produced in any lengths, breadths, and thicknesses. To those who have never witnessed the easy manner in which iron is rolled, it is difficult to convey an adequate conception of the productive power of even one ordinary furnace and rolling mill. In a week a great many tons of iron can be finished, and the yield of consecutive weeks would soon provide the quantity entering into the construction of such gigantic ships as those of the *Minotaur* and the *Warrior* class. A fraction of the time required to season timber would suffice to take the ore from the bowels of the earth, and fashion the iron which it produced into a stately and practically invulnerable ironclad. In a word, less time, less labour, and less cost enter into the construction of iron than of wooden ships; and, while wood is extremely perishable, iron is practically imperishable. Obvious as all this is, it is too frequently overlooked—*The Dockyards and Shipyards of the Kingdom.*

Innovation unsanctioned by the Surveyors of the Navy and the Admiralty was scouted by the reputable and numerous class glorying in the name practical, and for a time indifferently welcomed by the few on whom the burning stigma of theorists always fastened. Blindly the majority resisted everything because the dockyards did so, and it was more the consequence of luck than good management that the British mercantile marine received its great development. A minority of theorists meanwhile struggled manfully. Authority, such as it was, did not charm them, and year after year witnessed the construction of iron ships on the Thames and Mersey, the Clyde and Tyne. Taking courage from success, the builders and owners of iron ships claimed privileges for them from which wooden ships were debarred, and iron shipyards soon after ceased to excite the wonder of those intellectually behind their age.

The labour revolution. The labour revolution was in due time consummated, and that antagonism created between the ship and dock yards which the Admiralty no doubt hope will long continue. The antiquated labour system of the wooden era would not transplant nor take root in the iron shipyards. A new order of things required to be created, and in exact proportion as this truth has been recognised and acted on has iron shipbuilding prospered. Those shipbuilders—and there have been many of them—with enough sagacity to lay aside old notions and strike out in an entirely new direction, have added greatly to their wealth and skill, while those without depth, originality, or enterprise may still be found, in near as well as remote districts, lamenting what has happened, and prophesying the speedy ruin of those whose arms seem to be extended further than is compatible with meeting payments for any lengthened period. It is not, of course, in the least strange that the Admiralty have these latter for their bed-fellows. But it must be a matter of regret and humiliation to all Englishmen that the professional advisers of the Admiralty are chiefly to be found among the lees and dregs of scientific progress. What was first of all necessary in establishing iron shipbuilding on a businesslike and enduring basis was precisely what Adam Smith would have recommended, namely—divided labour. Men were required, not to do anything and everything as in the wooden shipyards, but to devote themselves to one separate branch. There were furnace-men wanted and frame binders, angle-iron smiths, platers, riveters, drillers, clippers, caulkers, &c. But these classes were unknown, and their creation furnishes an instructive episode in the history of British industry. No apprenticeships were served, and no bar existed or was formed to the free entry of the employed. The iron shipyards were opened to intelligent but unskilled working men, and so well were these working men guided that before long an average rate of wages was established that has not been greatly exceeded since. This is a striking proof of aptitude, and a telling condemnation of the system of long apprenticeships which some still favour. Taking the hint from this successful organisation, the Admiralty set the wooden shipwrights to work on the iron *Achilles* in Chatham Dockyard, seemingly unconscious that by so doing they were dealing the death-blow to the mixed dockyard labour system that still is so fondly cherished in the dockyards. The Admiralty may not, however, know that wooden shipbuilding is as simple, as easily learned, and as capable of minutely divided labour as iron shipbuilding. That wooden shipbuilding has never been subdivided in this country is owing to the blind dependence so long reposed by the shipowners and shipbuilders in the dockyards, and if it is never now subdivided it will be because it is not worth the while of any one in the presence of the increasing popularity of iron ships. The subdivision of wooden shipbuilding labour in the dockyards implies the complete overthrow of the existing system, and the rooting out of almost all the scandalous abuses that now exist.

The American Shipyards. American shipyard labour exemplifies the practical working of perfectly divided wooden shipbuilding. It shows the operation of a system anterior to and identical with iron shipbuilding as at present most successfully carried on in the great iron shipyards of this country. When a ship, be it of war or commerce, is contracted for in one of the great iron shipyards, the transaction presents itself in something of this form to the builder and the foreign Government or home merchant. The foreign Government or home merchant desires a particular class of ship to measure or to carry so many tons; and, given the price per ton, the whole outlay becomes an easy computation. What, then, is the price per ton, here, or there, and what special advantage will accrue from building the ship in one place rather than another, and by the hands of one firm rather than another? These are the questions for the foreign Government or the home merchant to consider. On the other side, the builder considers only what the ship will cost himself, and what in the particular case he ought to charge. What will it cost him to set up the frame of such a ship; what to plate such a ship; what to complete the inside work of such a ship; and what to supply the extras, such as spars, &c.? Mr. Brown is the foreman of all the framework, and he will by ten o'clock next morning furnish perfectly reliable statements of the cost of setting up the frame; if required, he will by that time also furnish contracts with the men to complete the frame for a certain sum. Mr. Black is the foreman of all the plate work; Mr. Green the foreman of all the inside work; and Mr. Grey the foreman of all the extras. These gentlemen are equally ready with Mr. Brown by ten o'clock next morning, and when the agent of the foreign Government or the home merchant calls again, the builder is prepared to treat. He knows exactly what the ship will cost himself, and the point for him to consider is, how much in excess of that amount is he to charge to receive the job? Assuming that he and his customer agree, the foreman of each of the working stages will almost to a week calculate when their men will finish, working the usual number of hours daily, and when they will finish working so many extra hours daily. Such is the admirable simplicity and perfection of iron shipbuilding as carried on in the great iron shipyards, and in many of the minor ones as well. So it has long been with American wooden shipbuilding. Of one of the well-known establishments in New York it has been said that the staff consists of the principal, one book-keeper, and one foreman. He (the principal) and his two assistants were, and no doubt still are, equal to the task of turning out 30,000 tons of shipping between the spring and autumn. The mode of operation is something of this kind. A ship of war, say the *General Admiral*, is wanted for the Russian Government, or a clipper for the New York and China trade. What either or both will cost, the principal will almost tell off-hand. But it may happen that the labour market is at the moment agitated from antipathy to Irishmen or negroes, and he can only let you know to-morrow. He in that case repairs to his office, consults his foreman and book-keeper; the calculations are entered into, and the boss foremen are hunted up in the back slums of the Bowery. These boss foremen are informed of the job, and each is asked to name his price and the time the work will take. It is then the turn of the boss foremen to look about and get together enough men to perform the work. The men once together, think together, and make up their minds both as to price and time. The principal is now in the position of the iron shipbuilder He knows, not doubtfully, but positively, what the *General Admiral* or the China clipper ship will cost. If he receives the job, he and his foreman lay out the work. The book-keeper keeps the accounts of the incomings and outgoings of the yard, and when the ship or ships are finished, the boss foremen are ready for another visit from any other New York shipbuilder.

<small>New York experience of an English Dockyard Officer.</small> An English dockyard officer, soured with the treatment he was receiving, bid adieu to the service some three or four years ago, and found his way first to New York, afterwards to the Lakes, and subsequently to the Mississippi river, from which he returned home before the outbreak of the war. He sought work in some of the New York shipyards, and as he could put his hand to anything, like all other English dockyard men, he was soon employed. One of the boss foremen hired him to make rudders, and he began in the most approved dockyard style. At the end of a week of hard work he repaired to the office

and found to his mortification that of all the men to be paid he had the smallest sum to receive. Others making rudders had earned twice as much, and some even more, and yet he, a young fellow, some time out of a full and long dockyard apprenticeship, and reckoned one of the crack dockyard hands, had done his best. What was the matter? How was it that these wiry Yankees, who seemed to take things so comfortably, thus mastered him? Time soon showed. The rudder makers earning twice as much were rudder makers. That was their business. They did nothing else. They therefore knew how to make rudders, and the English dockyard officer did not. Nor could the English dockyard officer learn by mere looking on, any more than one by merely looking at the performance of a shorthand-writer can learn shorthand. There were principles as well as knack to be acquired in making rudders as well as there is in making pins. By and by, through good-fellowship, coaxing, and bribing, the English dockyard officer mastered the business of making rudders, and was recognised as a New York boss rudder maker. On one occasion among those he hired to make rudders was a broken-down English shipbuilder, still very well known by reputation in this country. The tongue revealed the kindred of hirer and hired, and the latter frankly expressed his sympathy for the former. He had no faith in the job paying, having been himself bitten as a boss rudder maker. "Never mind me," remarked the English dockyard officer; "do exactly as I tell you, ask no questions, and if the job is finished on Saturday afternoon I will throw in an extra ten-dollar bill to your share." Saturday night came, the job was finished, and the English dockyard officer could well afford to give the promised ten-dollar bill gratuity to the broken-down

English shipbuilder. The English dockyard officer had learned to make rudders, knew what he was about, while the broken-down English shipbuilder was as dark, conceited, and incredulous as ever.

<small>Work on the American Lakes and Mississippi.</small> On the American lakes, as in New York, the English dockyard officer found wooden ship work divided into numerous separate and distinct branches, and farmed by boss foremen. There were boss converters of timber for ships' frames, boss fitters of ships' frames, boss deck and plank converters, and boss deck and plank layers, boss hold and cabin fitters, boss riggers, &c. In one and all of these trades he, the crack dockyard hand, was as much behind as before in the making of ships' rudders. He was an apprentice by the side of journeymen. Nay, he was worse, because the old-fashioned, slow, imperfect, and wasteful English dockyard way of doing everything had to be unlearned. He had to confront the principles and the knack that constant application in the doing of one thing had in the one case given and in the other inductively treasured up. It is a simple-looking thing to convert the timbers of a ship, and so it seems to turn somersaults and perform feats with cups and balls. But to do either to perfection, principles have consciously or unconsciously to be formed and applied. Even walking and running admit of cultivation. Why people do not seek to become great runners and great walkers is because there is no inducement offered to them; and why wooden ship work in the English dockyards remains at low-water mark is because English dockyard shipwrights have not been drawn out as the American shipwrights have been. The demand for ships in America in a month or two, or in three or six

months or not at all, and at a price beyond which not another farthing will be paid, has thrown the shipbuilders into the arms of the working men. The shipbuilder has become a middleman between the workmen and the shipowner, in much the same sense as an architect is a middleman between the capitalist who wants a house, and the different tradesmen—the bricklayer, the carpenter, the plumber, the gasfitter, &c.—who are to perform the work. An architect will tell you what the house you want will cost, and if you were to ask him to provide it for the sum he names his position would be identical with that of the American shipbuilder. The architect would then contract with the various tradesmen and the work proceed. But houses are not built that way in this country, and the building trades are still unbroken by the calling together of boss foremen and their men to know what they will do this and that for in a given time. In American shipbuilding this, however, has been done, and the result is that classes of men have settled down to the doing of certain things which the potent encouragement of money has learned them to do so well, that an English dockyard shipwright is not entitled to the name of tradesman in their presence.

The harmony of contract interests. In this divided contract labour system there is a complete harmony of interests throughout. The practical shipowner knows what he can pay for a ship, and he agrees to pay it. What more has he to think of or care about in the transaction? The specifications are in his drawer, the work is supervised either by himself or by one in whom he has full confidence, and at each of the paying stages of the ship compliance or non-compliance with the contract is visible at a glance. The contracting for a ship is therefore, as regards the shipowner, as satisfactory as the contracting for any other thing. No doubt a deal of fuss is often made, but it arises usually from business ignorance. A thorough man-of-business shipowner has no more reason to mistrust a shipbuilder than he has his banker, his charterer, or his broker. Then take the case of the shipbuilder. He has entered into a transaction with his eyes open. He has agreed to build a ship of a certain form and tonnage, with this and that kind of timber in different parts, by a certain time. There should be nothing for him to grumble at. The ship should remunerate him, and if he is a man of business it will remunerate him. He has sufficient knowledge of the world to know that the owner is a safe and good man, and enough sagacity to turn his hand, lose nothing, and maintain position and credit should the owner fail. Among the shipbuilders of this country there are many men of this stamp, and no doubt there are others below the mark. Of the latter there is no use to speak, for they deserve only to be weeded out. Then take the case of the workmen. They know their work, and they know each other. Among themselves they have learned—in a rude way it may be, but still a most efficient one—to estimate what they can do in a day, a week, a month, a year. The job they have in hand is to pay them well if completed within the time, but not otherwise. In case of failure, there will be no bonus, and another job waiting elsewhere may be lost. With nothing, therefore, but their own thews and sinews have they fault to find: and these never fail. Go from the workmen to the shipbuilder, and from the latter to the shipowner, there is perfect harmony. The shipowner will also often feel himself able, particularly in good times, to stretch a point in price for the ship he wants, and so will the shipbuilder as regards the workmen, if he desires to get other work in hand that is waiting for him. Of both advantages the workmen are the recipients, and who deserve it better? Paying workmen well is a practice of which none need be the least afraid, and contract work is of all ways the best calculated to develope their energy and skill.

Applicability of the system to the Dockyards. That the system of divided contract labour is applicable to the dockyards will now hardly be disputed. But if it is, then let it be observed that in America, throughout the war, it has been subjected to an entirely new and severe trial. The great fleets and flotillas that have so recently been created are chiefly if not altogether the fruit of this labour system. America of late has built its ships of war by contract. Naval officers and others have prepared models, drawings, specifications, &c., and these being sanctioned by the Secretary for the Navy, the shipbuilding market has been entered in much the same way for ships of war as it usually is for others. New York, Boston, Philadelphia, &c., were invited to say what they could do, and the boss foremen and the workmen were consulted as to time and price. If hitches have occurred at all, they have been of extremely rare occurrence. Ships of war have been produced as if by enchantment, and some of the Monitors were only a few weeks in hand. But for its system of divided contract labour the production of ships of war in America would have been as slow a process as with ourselves. Surely, then, such a system is applicable to the public service! Surely it is the system that is alone applicable both for repairing and constructing ships of war! By its introduction the first thing to be determined would be the sum that either repairs or a ship would cost. That calculation, resting mainly on the price charged by the workmen for the different stages of the work would give the dockyard workmen a standing that they do not now possess, and increase their earnings without necessarily imposing new burdens on the Exchequer and the public. It would elevate them to the condition of the workmen in our private shipyards, and render superannuation and all the other forms of pauperism supererogatory. When a ship is contracted for in the private shipyards and afterwards contracted for by the workmen, the practice is to pay agreed-on weekly wages until the different stages of the work are in turn completed. Let us say £1,000 is to be paid for the first stage of the work, and that at its completion the wages paid amount to £700 or £800, then, as it happens, £300 or £200 is disbursed as a bonus among the workmen. So with the other stages. The system, of course, works admirably, and one and all have a strong motive to exertion. In the dockyards there is none of this, for when a workman earns more than the limit of his earnings he gets nothing for his pains, while if he earns less, he is only paid what he really earns. Change, therefore, is required, and whether it is to the practice of our own shipyards or to the practice of the American, Parliament some day must decide. Either is applicable, but that manifestly the more which divides labour thoroughly, and holds out to thoroughly divided labour its full and relatively great reward. The next movement of the British shipbuilders must be in the direction of the American system, and those charged with the responsibility, such as it is, of the dockyards, should not deal by halves with a question in which all classes have so deep an interest.

Chapter III.

DOCKYARD MANUFACTURES.

<small>Sir George Cornewall Lewis's principle of manufactures.</small> The principle that should govern manufactures in public establishments was forcibly laid down in one of the last speeches of the late Sir George Cornewall Lewis. On the 23rd March, 1863, he said, "With regard to the general question of Government manufactories, an article which may properly be produced by contract must fulfil two conditions: the first, that the article should be capable of verification; and the second, that it should be of a sort for which there is a natural demand by private persons. There could, for instance, be no justification for the Government setting up a cloth manufactory." In other words, nothing should be manufactured in the dockyards that can be purchased, and the quality of which can be tested. The rule is unexceptionable, because, as remarked by Sir George, there are commodities—particular makes of cannon, for example—which the public do not want, and which it may be advantageous for the country to possess; and because the great bulk of the commodities required for the public service admit of easy and complete verification, both as regards quantity and quality. Take the case of cloth, for the manufacture of which Sir George said Government could find no justification. The quality of the cloth would be disclosed when examined by a moderately powerful glass. And the quantity of the cloth

can be measured. Last of all, cloth is in regular demand, is produced by competing manufacturers, and sold by competing shopkeepers; therefore cloth bought in the open market is obtained on the most advantageous terms possible. On the other hand, were the quality of cloth difficult or impossible to determine, were the quantity delivered incapable of weight or measurement, or, finally, were the production the close monopoly of an individual or a class, then cloth should perhaps be manufactured in public establishments for the public service.

<small>The application of this principle to the Army.</small> How far the application of this principle is carried out by the department over which Sir George Cornewall Lewis last presided may be vaguely gathered from a return, dated the 15th April, 1863, and attested by Earl De Grey and Ripon.* The first in order of the army manufacturing establishments is the Royal Carriage Department, Woolwich Arsenal. What is called its production voucher embraces several pages of printed items; but it is a "return of stores manufactured, and of stores obtained direct by contract, as well as of stores delivered from contractors to the principal superintendent of stores." The stores manufactured, the stores obtained by contract, and the stores delivered, are not distinguished; they are alphabetically grouped, and their aggregate value is stated at £329,008 6s. A separate line gives the value of the delivered stores included in the £329,008 6s., which is stated at £46,388 17s. 1d. The second in order of the army manufacturing establishments is the Royal Gun Factories, Woolwich Arsenal. Their production voucher embraces even more pages than the

---

* Return Army (Manufacturing Establishments); official number 176; price 9d.

previous voucher, but there are no contract items, unless those of the Armstrong guns supplied and the services rendered by the Elswick Ordnance Company. The third in order of the army manufacturing establishments is the Royal Laboratory, Woolwich Arsenal. Its production voucher fills no fewer than thirty pages, and in the issues to the principal superintendent of stores there is again no distinction between manufactured and contract articles; the alphabetical classification being still adhered to. The fourth in order of the army manufacturing establishments is the Royal Small Arms Factory, Enfield. The fifth and last in order of the army manufacturing establishments is the Royal Gunpowder Factory, Waltham Abbey. In the army, then, the principle of supply by contract is affirmed, at least verbally, as well as the principle of independent manufacture. Guns, rifles, and gunpowder are manufactured because, in a certain sense, they are not in natural demand; that is to say, in a certain sense there is not an open market demand for them. On the other hand, turning to the production voucher of the carriage department, it may be presumed that bags, barrels, bars, blocks, boilers, bungs, buckles, &c., are supplied by contract, because there is a natural demand for them, and quantity and quality admit of being verified. No doubt there are many flagrant departures from principle in the supply of articles to the army, but the manufacturing establishments, at least, possess the merit of being few in number.

<small>Enfield.</small> While, however, saying this much in commendation, the cases of Enfield and Elswick must not be passed over, these being instructive examples of official meddling and mismanagement. The case

of the Enfield small arms factory is also remarkable, inasmuch as the arguments advanced in support both of the establishment of that factory and afterwards for its extension are word for word identical with those advanced by the Admiralty in support of iron-cased shipbuilding at Chatham. Lord Raglan, before the 1854 Small Arms Committee, stated it to be unsafe to rely on the private trade, and that the Government factory would effect a large annual saving. Colonel Tulloch, before the same Committee, stated that no dependence could be placed on the private trade, and that Government had no protection against the exorbitant demands of the trade. Sir Thomas Hastings, before the same Committee, stated that the Birmingham gunmakers had completely broken down in their contracts, and therefore that it would be unsafe for the Government to place reliance on the gunmakers. Now, what did Admiral Robinson, the Controller of the Navy, say the other day against the private shipbuilders, and in support of iron-cased shipbuilding in the dockyards? He said, "In no one instance have the contractors kept to their agreements with the Government, either as to *time* or *cost*. The *Warrior* was ordered on the 11th May, 1859; the contractors agreed to deliver her complete in July, 1860. She was delivered on the 20th September, 1861, and not quite completed till the 24th October, 1861. The contractors agreed to build her for £210,225; including extras. They claimed, and ultimately received, £254,728. The *Black Prince* was ordered on the 6th October, 1859. The contractors agreed to deliver her complete on the 10th October, 1860. She was delivered incomplete on the 18th November, 1861. She was not completed for some months afterwards by the artificers of the dockyard. The contractors agreed to build her for £230,254. They have claimed for her £259,751, and part of this claim is still under consideration; she has, however, cost the Admiralty up to this time £249,751. The *Defence* was delivered incomplete four months after the contractors' agreement. The *Resistance* was delivered incomplete ten months after the contractors' agreement. The *Hector*, if delivered in the month of March, will be seven months behind the time agreed upon. The *Valiant* should have been delivered, according to the original contract, in August, 1862: the contractors failed, and requested the Admiralty to annul their agreement. Another contract was made with a fresh party, to deliver the ship in March, 1863: there is no prospect whatever of the work being finished for six months after that date. The same or even greater delay has taken place in the delivery of the *Orontes* and *Tamar*, although these ships are of simple construction and are not armour-plated. There is no prospect that more than one out of the four iron ships last ordered will be delivered till many months after the periods agreed upon. It is not, therefore, *one* contractor or *one* iron shipbuilder, but *all*, who have failed in their agreements; and this clearly indicates the great uncertainty attending this mode of construction. Two other difficulties present themselves. 1. The great* slovenliness

---

* Admiral Robinson charges the shipbuilders with unskilfulness, want of honesty, and with being lazy. Now, who is Admiral Robinson, that hazards these grave charges? What does he know of skill or want of skill in iron shipbuilding, or even in wooden shipbuilding? It is not so very long since Sir Baldwin Walker evaded the House committee by weighing anchor and making off to the Cape station with the *Narcissus;* and not till after Sir Baldwin's disappearance was Admiral Robinson known to fame as the Controller of the Navy. The Admiral may be ashamed to hear the truth, although I do not know why he should; but no sooner was he appointed to his present post for five years by the Duke of Somerset, than he went to the yard of the Thames Iron Shipbuilding Company to gather all the information that he could.

of the work performed by iron shipbuilders, rendering the presence of an Admiralty inspector necessary on the premises wherever the contract ships are building, and leading to many difficulties between the contractors and the Admiralty; and, 2. The great temptations that beset the contractors, owing to the cost and difficulty of procuring good iron, to use inferior and cheap material. Again, after a contract is signed, no alteration or improvement, however great, can be made without submitting to any terms the contractor chooses to enforce."† Very oddly, too, the reply of the Birmingham gunmakers is word for word identical with the replies of the iron shipbuilders. One and all of the charges against the gunmakers were met satisfactorily, and it was actually proved that the difficulty of obtaining a sufficient supply of small arms was occasioned by the Board of

---

Nor was he content to increase his own scanty stock of knowledge, but squad after squad of dockyard learners were sent by him to the same yard, that as many of the dockyard people as possible might become acquainted with the principles and practice of iron ship construction. Now Admiral Robinson, and the men who so very recently were learners, turn up their noses at their instructors and call them names. They presume to know right and wrong in iron ship construction, better than those who were practical iron shipbuilders many years ago.

And who, let me ask, are the workmen in iron in the dockyards who so far surpass the workmen in iron in the shipyards? They are the superseded wood joiners, caulkers, &c., of the dockyard towns, who never in their lives, until the other day, heated a bolt or made good a rivet. These men, says the Controller, are the only ones to trust to, because their workmanship is the best, because their honesty is not to be doubted (although the police search them when they leave work), and because when the Controller is near them they are hard-working. I cannot see the force of this. I say, it is impossible that the squads who in turn visited the Thames Shipbuilding Company's works so very recently, to learn the use of iron tools and the modes of doing iron work, can be anything but bunglers and unreliable workmen for a long period. I warn Parliament that the workmanship of the *Achilles*, in Chatham Dockyard, is not to be depended on; for the workmen are not proper tradesmen, and the Controller of the Navy is not entitled to be heard on the subject of the construction of iron ships. Parliament might as well listen to the inveterate advertiser of well-made trousers in Brook-street.—*The Dockyards and Shipyards of the Kingdom.*

† Navy; statement relative to the advantages of wood and iron, &c.; 3rd March, 1863. Price 3d.

Ordnance and the officers acting under it, as well as by delay in selecting patterns, the severity of the "view," and the enormous per-centage of rejection without proper cause. They also proved the capability of the trade to supply the wants of the Government, and to provide for a steadily increasing export demand. The rejection of suitable iron, the severity of the test for armour-plates, the changes of plan, and the delay in deciding on changes, account also for all the apparent shortcomings of the iron shipbuilders. But let us now turn to the lesson that the extension of the Enfield small arms factory teaches, that the present danger of iron shipbuilding in the dockyards may be perceived. Mr. Monsell and Mr. Gladstone recommended the extension of the factory, first, because the entire outlay would be no more than £150,000; and, second, that the saving consequent on supplying the needful weapons in that manner would effect an immediate saving of almost £1,000,000. The estimated saving was calculated in this way: 485,000 rifles at £3 each would by contract cost £1,455,000, and by manufacture £1 10s. each, thereby saving £727,500; and 485,000 bayonets at 7s. 6d. each by contract cost £181,875, and by manufacture 1s. 6d. each, thereby saving £145,500. Now, what the amount of money actually spent at Enfield has been, perhaps none can tell; and what the cost of the rifles and the ramrods has been, certainly none can tell. What alone is clear is that the statements and calculations put forward by the Government were worthless and untrue, and that the public, in addition to being better served by the gunmakers, would have been spared the sheer waste of very large sums. And not one of the least of the evils of the extension of the Enfield factory was the unsettling of a great branch of domestic industry, to the serious loss as well as inconvenience both of employer and employed.

<small>Elswick.</small> The report of the Select Committee on Ordnance sets the Elswick transaction in its true light.* The Government began the manufacture of Armstrong guns, and £2,539,547 17s. 8d. has literally been thrown away, there being only a few guns of doubtful utility to show for this very large sum. Of this sum £965,117 9s. 7d. has been paid to the Elswick Ordnance Company for articles supplied, £65,534 4s. as compensation for terminating the contract, and there are still outstanding obligations from the War-office to the company for £37,143 2s. 10d. The sum of £1,471,753 1s. 3d. has been expended in the three manufacturing departments at Woolwich, on Armstrong guns, ammunition, and carriages. Those articles which were manufactured both at Elswick and Woolwich, and which therefore are supposed to admit of direct comparison, would, according to Mr. Whiffin, if all manufactured at Woolwich, have led to a saving of £242,173 10s. 6d. on an expenditure of £593,275 10s. 11d. In other words, the same articles cost twice as much at the one place as at the other, although both were under the same control and supervision, and although Elswick had a considerable advantage in cheaper coal. But while Mr. Whiffin and Mr. Boxer, Superintendent of the Royal Laboratory, state that the comparison of prices is substantially correct, Mr. Baring, sometime Under-Secretary of State for War, and others, say that the comparison is of no value for practical

---

* Report from the Select Committee on Ordnance; July, 1863. Price 4d.

purposes. No doubt Mr. Baring knows best, and the 1859 Dockyard Committee in their report certainly direct attention to the often-practised facility that exists of creating deficiencies and excesses by merely complicating the entries a little more than they are. No doubt Mr. Baring distrusts all official figures. The 12-pounder Armstrong field pieces are believed by the Committee to be efficient. Of the 40-pounders they do not say much, but the 110-pounder they regard as a valuable addition to a ship's armament for chasing or running away purposes. Still, notwithstanding the expenditure of £2,539,547 17s. 8d. on Armstrong guns, the Committee formally conclude, "The old 68-pounder is the most effective gun in the service against iron plates." This is indeed melancholy. Mr. Armstrong created a Baronet by mistake, and two millions and a half of the public money wasted by mistake! But to what other conclusion could the Select Committee have come? They must have known the conditions under which the firing of big guns takes place at Shoeburyness. No one is exposed. The artillerymen scamper off to the splinter-proofs, and after a subaltern is satisfied that no person is within reach the bugle sounds the fire, and from a remote splinter-proof the firing string is drawn. Such a system may be admirably adapted to test armour-plates, but why any practical value has been placed on guns fired in such a manner baffles comprehension. The Committee, therefore, did quite right in falling back on the 68-pounder as the only really serviceable gun for tough work that we yet possess.

The Dockyards governed by no rule. The dockyards acknowledge no manufacturing rule. No Sir George Lewis has yet presided over them. Within them are grouped together the innovations, prejudices, and *débris* of ages. No reform worthy of the name has, since the establishment of the dockyards, been introduced, and these establishments are therefore a standing protest against system as well as a sanction to absurdity and waste.

One of the Dockyard objections to contract supplies. One of the dockyard objections to contract supplies, and consequently one of the reasons for some part of the confusion that prevails, is to be found in the report of the 1859 Committee. It appears that in 1834 the dockyard authorities by some means or other managed to establish fair standard prices for ironmongery, nails, tools, and various other articles in what is known as the hardware line. Pattern rooms were founded, arranged as fantastically as the stocks, barrels, locks, and ramrods in the Tower of London, and to each article a parchment Prince of Wales' feather revealed with little effort to the lazy clerks the just price as between the Admiralty and the manufacturer. The task no doubt was a great one, and when finished a grateful feeling of relief could scarcely fail to pervade official circles. But will it be believed that for more than twenty years there was no recovery from the price and pattern effort? In 1859 the Committee of that year found the pattern rooms precisely as when first established. Her Majesty's artificers and the fleet were supplied with obsolete hardware at obsolete prices. The refuse of over-stocked colonial and foreign markets and the stocks of bankrupt country ironmongers had for years been poured into the Royal Dockyards and paid for at the extravagant rates current many years before. The dockyard authorities were of course ignorant of any change, and not aware that they were under any obligation to see to the public interest in such things. They had special duties to perform, and it was not within their province to know that over twenty years had all but revolutionised the hardware trade, introducing improvements without number, and materially diminishing the previous prices. They were entirely guiltless. If blame rested anywhere, it was on those who profited by supplying old-fashioned articles at old-fashioned prices. And it was added that the old-fashioned hardware was only one of many proofs that no faith whatever could be reposed in contractors, all the honesty of the country subsisting quietly and decently on the produce of the public burdens inside the public dockyards. It is thus an objection to contract-supplies in the dockyards that the worse than Turkish apathy of the dockyard authorities has to be aroused for the discharge of a common-sense and proper duty.

One of the Dockyard objections to contract shipbuilding. Again, one of the dockyard objections to contract shipbuilding is to be found in the report of the Committee on Gun and Mortar Boats.* The gun and mortar boats were built in 1854, 1855, and 1856, of such timber as could be found, by such men as could be got, and with all possible despatch. Of course a house built under such conditions would be in danger of tumbling down before it was completely finished, and the gun and mortar boats were necessarily, in the main, an indifferent job. On this point, however, there is no doubt, for it was given in evidence by Sir Baldwin Walker, on the authority of Sir James Graham, that so great was the pressure for the gun and mortar boats that there could be no standing on stepping-stones

---

* Report Navy (Gun and Mortar Boats); 1860. Price 2s. 6d.

as to timber. Sir Maurice Berkeley said the same thing. Still, in the face of this testimony, and in the face of the circumstances with which Sir Charles Napier made every one acquainted, the imperfections of the gun and mortar boats are to this day charged against the shipbuilders, as proof of professional incompetency and a disposition to indulge in roguery. Explanation is unheeded, and the fact ignored that the gun and mortar boats were built under the inspection of dockyard officers. It is enough for the dockyard authorities that indifferent gun and mortar boats were built by the shipbuilders for the public service. Surely a more narrow and disingenuous judgment could not possibly be passed. It as thoroughly betrays the spirit of prejudice and unworthiness that actuates the dockyard authorities, as the discovery of an unaltered hardware contract of more than twenty years' standing betrays unfitness for the positions that they at present fill.*

The Mast-houses.† But the unfitness and prejudices of the dockyard officers and the inconsistencies that pervade the dockyards stand in no need of the proof that might be furnished by grouping together objections of various kinds and tracing each to the causes that have just been named. The more satisfactory course will be to pass in review the various manufacturing establishments which the Admiralty at the present time main-

---

* The 1859 Committee found not only the timbers of the gunboats unventilated, but the holds unventilated, so that the entire woodwork under a burning sun on the slip at Haslar constantly underwent the trying ordeal of a baker's oven.

† The Committee found the mast and boat houses generally in a very unsatisfactory state. They were used too much as places of store and deposit for made masts and boats in a complete state, so as to deprive the men of proper room to work, and prevent the supervising officers from being able to see their men at work.—Page 46, Report of the 1859 Committee.

DOCKYARD MANUFACTURES.

tain.* First in order stand the mast-houses. These are the houses in which masts are made, repaired, and stored, and during the year 1860-61 the expenditure in these houses amounted to no less than £110,042 15s. 9¼d. At Deptford there was an expenditure of £178 7s. 3d., inclusive of £72 0s. 7d. for material, and £88 3s. for labour; at Woolwich there was an expenditure of £11,221 2s. 3d., inclusive of £8,469 7s. 2d. for material, and £1,845 13s. 7d. for labour; at Chatham there was an expenditure of £20,927 10s. 8¾d., inclusive of £15,775 14s. 9d. for material, and £2,614 8s. 1½d. for labour; at Sheerness there was an expenditure of £13,552 16s. 8d., inclusive of £9,503 11s. 9¼d. for material, and £2,250 6s. 3¼d. for labour; at Portsmouth there was an expenditure of £44,051 8s. 4d., inclusive of £33,603 3s. 1d. for material, and £5,530 9s. for labour; at Devonport there was an expenditure of £20,111 10s. 6½d., inclusive of £11,778 2s. 6¾d. for material, and £3,043 3s. 5¾d. for labour. Let us, further, turn for a moment to the credit side of one of the accounts; say the Portsmouth account, it being the largest. In Portsmouth Dockyard during the year 1860-61 there were 499 masts manufactured, and what may be called proportionate numbers of bowsprits, main yards, topsail yards, booms, jib-booms, gaffs, &c., in addition to 333 boats' masts, 161 boats' yards, &c. Altogether in the one year 1860-61 the dockyard masthouses seem to have turned out a sufficient number of spars to rig the fleets of the whole world, and certainly as many spars as the British Navy will wear out in twenty years. This is a startling fact; but what is to be thought of it when it is added that the dockyard mast-houses are

---

* Balance-sheets showing the cost of manufacturing articles in the workshops of the several dockyards and steam factories for the year 1860-61; 1862. Price 4s.

still in full swing, and have been doing a roaring trade since the peace in 1816? To provide storage for masts and spars of course explains to some inconsiderable extent the necessity that exists for new dockyard works, dockyard extensions, and other familiar but really monstrous phrases.

<small>The Boat-houses.</small> Second in order stand the boat-houses. These are the houses in which boats are made, repaired, and stored, and during the year 1860-61 the expenditure in these houses amounted to £23,824 17s. 5¼d. At Woolwich there was an expenditure of £2,239 3s. 6d., inclusive of £1,007 16s. 6d. for material, and £871 14s. 10d. for labour; at Chatham there was an expenditure of £6,837 18s. 8d., inclusive of £3,446 14s. for material, and £2,488 3s. 9¼d. for labour; at Sheerness there was an expenditure of £2,223 19s. 9d., inclusive of £849 6s. 11½d. for material, and £1,106 13s. 8¾d. for labour; at Portsmouth there was an expenditure of £7,172 4s. 3d., inclusive of £3,359 19s. 6d. for material, and £2,774 15s. 10d. for labour; at Devonport there was an expenditure of £5,351 19s. 3d., inclusive of £1,186 12s. 4¾d. for material, and £797 16s. 11d. for labour. Let us again turn to the credit side of the account for Portsmouth. In Portsmouth Dockyard during the year 1860-61 there were 130 boats built or fitted. And in Devonport Dockyard, in addition to the boats building, there were 123 boats fitted for store, 55 boats refitted for store, 53 boats repairing for store, 109 boats repairing for harbour service, and 4 other boats in hand. The dockyard boat-houses were therefore well employed in 1860-61, and as boat building and repairing have also been proceeding without intermission for more than half a century, it

would not be a useless question for young debaters to discuss whether or no there are at present in the dockyards as many boats as would form a belt or bridge round the globe. One hundred and twenty-five launches placed stem and stern form as near as can be a line a statute mile in length.

The Capstan-houses. Third in order stand the capstan-houses. These are the houses in which capstans, anchor stocks, pump boxes, &c., are manufactured, repaired, and stored, and during the year 1860-61 the expenditure in these houses amounted to £10,857 0s. 10¼d. At Woolwich there was a total expenditure of £2,138 4s. 5d.; at Chatham of 3,782 6s. 8¾d.; at Sheerness of £757 2s. 3¼d.; at Portsmouth of £572 11s. 5d.; and at Devonport of £3,084 18s. 7d. Here manifestly in these capstan-houses is the maintenance of five small manufacturing establishments, with all their costly adjuncts. Here are capstan-houses because there were capstan-houses in the days of Blake and Nelson, and because there is a crop of people young and old attached to them, who in turn will be entitled to superannuation.

The Joiners' Shops.[*] Fourth in order stand the joiners' shops. During 1860-61 there was a total expenditure at Deptford under this head of £897 4s. 5d.; at Woolwich of £1,560 4s. 8d.; at Chatham of £2,677 4s. 10d.; at Sheerness of £3,639 10s. 8d.; at Portsmouth of £3,545 4s. 3d.; at Devonport of £3,976 4s. 7½d.; and

---

[*] The Committee are of opinion that sufficient appreciation has not been given to the advantage of introducing machinery generally in the joiners' department. In some shops steam-power to drive the machines has not yet been introduced, and to this subject the Committee consider that immediate attention ought to be given in every instance in which arrangements to do so have not been already made.—Page 49, Report of the 1859 Committee.

at Pembroke of £1,295 12s. 1d. Aggregate, £17,591 5s. 6¼d. Turning to the credit side of the Portsmouth account, an insight is obtained into the kind of work performed by the dockyard joiners. The items in the account, the numbers of which exceed 100, are the following:—Boxes, cartouche, fitted for seamen, 213; boxes, fir, for seamen's kit, 187; cot frames, 251; roller blinds, complete, 197; tables, mahogany, hanging, 378; tables, mahogany, telescope, dining, No. 4, 321 feet; ditto, iron legs, No. 9, 974 feet; ditto, ledged, No. 7, 273 feet, and bed frames, iron, repairs, 119. The chief employment of the dockyard joiners is making dining-tables, and their annual production of that very useful article measures very nearly threequarters of a mile. How many miles of dining-tables are at the present time in store in the dockyards? That dining-tables, roller blinds, cot frames, and boxes would be much better manufactured out of the dockyards than in them, few will doubt, and it will astonish most people that a class of workmen who were supposed to have a great deal at stake in the continuance of wooden shipbuilding, have in reality nothing to do with the hulls or spars of ships, whether these are of wood or iron.

The Plumbers' Shops. Fifth in order stand the plumbers' shops. The total expenditure for the year 1860-61 under that head was £4,765 18s. 8¼d. Woolwich expended £64 12s. 3d., inclusive of £46 10s. 2d. for material, and £12 5s. 10d. for labour. Chatham expended £3,976 15s. 11¾d., inclusive of £3,058 18s. 5¼d. for material, and £467 14s. 1d. for labour. Sheerness expended £103 14s. 10½d., inclusive of £51 7s. 1¾d. for material, and £51 7s. 1¾d. for labour. Portsmouth expended £53 6s. 11d., inclusive of £40 3s. 7d. for material, and £6 11s. 10d. for labour. Devonport expended £567 8s. 8d., inclusive of £207 1s. 3¾d. for material, and £290 2s. 9d. for labour. Turning to the credit side of the Portsmouth account, there were during the year 50 lead inkstands manufactured, 200 fishing leads, 400 hand leads, 9 lead scuppers, and 4 lead soles for boots. These are the whole items for the year. Lead inkstands no doubt for the convict warders, lead sinkers for dockyard gentlemen fond of ground fishing, and two pairs of lead soles for some one's boots. Than the plumbers' shops there is surely not a greater conceivable abuse. An establishment in Woolwich Dockyard paying £12 5s. 10d. per annum in wages, and a similar establishment in Portsmouth Dockyard paying £6 11s. 10d. per annum, are inscrutable. The only probable conjecture is that the accounts are purposely kept low.

The Wheelwrights' Shops.*   Sixth in order stand the wheelwrights' shops. The charge for these for the year was £1,827 5s. 10¼d. Deptford expended £82 6s. 11d., inclusive of £22 17s. 10d. for material, and £12 1s. 7d. for labour; Woolwich, £97 13s. 11d., inclusive of £57 14s. for material, and £25 19s. 7d. for labour; Chatham, £198 14s. 6½d., inclusive of £118 14s. 10d. for material, and £61 0s. 7d. for labour; Sheerness, £608 3s. 11¾d., inclusive of £396 0s. 8d. for material, and £145 10s. 11d. for labour; Portsmouth, £715 16s. 7d., inclusive of £380 0s. 2d. for material, and £195 17s. 9d. for labour; and Devonport, £124

---

* With reference to the wheelwrights' shops, the Committee found that in all the yards wheels had not received that amount of attention that was desirable. Cabinetmakers and joiners seem hitherto to have been placed upon wheel work indiscriminately with the work for the bodies of the carts and waggons, which is objectionable, as leading to the production of inferior wheels, and should be discontinued.—Page 50, Report of the 1859 Committee.

9s. 11¼d., inclusive of £71 7s. 11d. for material, and £38 8s. 2d. for labour. Turning to the credit side of the Portsmouth account, there were 24 balers made for boats, 24 targets for musket practice, 24 trestles for sawyers, 25 grindstone troughs, 42 rollers for sawyers, 46 chests (tool or store), and 288 oak bars for racks. These were the great items; the small items embrace lamplighters' ladders, &c. Comment surely is unnecessary. Wheelwrights' shops in the dockyards, because in the time of Henry VIII. some ships were made on wheels, is monstrous. And to think also to what base purposes this class of artisans has been turned! Old officials approaching the happy confines of superannuation must view the wheelwrights' shops with emotion—repositories of grindstone troughs and water-balers for boats.

The Millwrights' Shops.*   Seventh in order stand the millwrights' shops. The charge for these for the year was £17,081 0s. 10d. Chatham expended £5,774 3s. 10½d., inclusive of £4,251 3s. 11¾d. for material, and £851 14s. 8¼d. for labour; Sheerness £322 0s. 4¼d., inclusive of £120 17s. 8¼d. for material, and £125 4s. 11d. for labour; Portsmouth, in all, £6,389 5s. 2d.; and Devonport, £4,595 11s. 5½d., inclusive of £2,474 18s. 2d. for material, and £1,253 17s. 3¼d. for labour. Turning to the credit side of the Portsmouth account, there were blocks, rings, chains, nuts, plates, pins, screws, tallies, tillers, and "conversions" for the money; but turning to the credit side of the Devonport account, there were in addition capstan apparatus—

---

* The Committee are of opinion that the dockyard millwright shops should be abolished in all the yards where there is a factory, and that the work hitherto done therein should be done in the factory, or under factory supervision.—Page 57, Report of the 1859 Committee.

another capstan-house, in fact. Few will find fault with this sum, for it is much less than all might expect. However, by millwrights' shops one usually understands something quite different from another dockyard capstan-house, and shops in which blocks are made, although there are blockmakers' shops in the dockyards, and shops in which rings, chains, nuts, plates, &c., are manufactured, although there are smitheries and engineers' shops in the dockyards. The dockyard millwright shops, like the wheelwrights' shops, the plumbers' shops, the joiners' shops, and the capstan-houses, should be at once gutted and demolished. No doubt there were good reasons for establishing these shops; now there are equally good reasons for their being forthwith rooted out.

The Roperies.* Eighth in order stand the roperies, although they are first in importance by rather more than £100,000. The expenditure in the roperies for the year 1860-61 was no less than £371,671 5s. 0½d. At Chatham 1,142 tons 13cwt. 3qr. 20lb. hemp were spun in the spinning-loft, at a cost of £45,056 4s. 7¾d.; 342 tons 7cwt. 2qrs. 14lb. hemp were spun at the spinning-machines, at a cost of £11,084 8s. 8d.; and the tarring-house returns were 1,452 tons 7cwt. 3qr. 21lb., at a cost of £52,610 8s. Then at Chatham there is a charge of £2,672 7s. 6d. for the preparation and cutting of hide thongs; £2,727 11s. 9¼d. for making strands and closing hide tiller ropes; £3,425 13s. 6½d. for unlaying worn cordage; and £75,187 13s. 4½d., is the charge for the laying-houses. The total expenditure at Chatham for material

---

* Notwithstanding the complaints of the master ropemakers in the several dockyards that the supervision is inefficient, the Committee are of opinion that by proper distribution of officers the work may be sufficiently supervised without an additional foreman as proposed by them. The Committee further recommend the employment of women in place of boys, as in the private trade.—Page 63, Report 1859 Committee.

was £116,318 10s. 0¾d., and labour £17,562 4s. 9½d. With the usual extras added the total reached £142,320 0s. 10d. At Sheerness the charge for material was £1,270 2s. 11d., and for labour £77 17s. With the usual charges in addition the total Sheerness expenditure was £1,421 2s. 3¾d. At Portsmouth in the spinning-lofts the charge for material was £35,809 8s., and for labour £8,748 8s. 3d.; in the tarring-houses the charge for material was £1,126 0s. 2d., and for labour £1,749 13s. 7d.; in the yarn selected from worn cordage the charge for material was £1,642 13s. 10d., and for labour £127 10s. 7d.; in the laying-houses the charge for material was £646 7s. 10d., and for labour £1,282 10s. 6d.; in laying white yarns into cordage the charge for material was £44 10s. 2d., and for labour £55 18s. 3d.; and in relaying manufactured cordage the charge for material was £136 18s., and the charge for labour £232 14s. 5d. In none of these sums are the usual charges for the share of general expenditure and percentage on the cost of produce included. The Devonport roperies are on the Chatham scale of magnitude, and it will suffice to say that at Devonport the expenditure for the year was no less than £169,281 3s. 0¾d. Whether rope should be manufactured in the dockyard is a question on which considerable difference of opinion will prevail, although it admits of the application of Sir George Lewis's test, both as regards quantity and quality. That, however, such gigantic establishments as the Chatham and Devonport roperies require close Parliamentary watching and scrutiny admits of no question; neither does the criminal—for it is nothing less—accumulation of rope far in excess of present or prospective wants, although we were to be at war with the whole world.*

---

* Next to bringing the shipyards and dockyards into dependence and harmony,

The Sail-lofts. Ninth in order stand the sail-lofts. The charge for these for the year was the considerable sum of £69,657 11s. 2¾d. At Deptford the expenditure was £9,523 7s. 11d., inclusive of £6,548 19s. 5d. for material, and £1,335 14s. 4d. for labour. At Woolwich the expenditure was £6,576 3s., inclusive of £4,727 15s. 1d. for material, and £982 8s. 8d. for labour. At Chatham the expenditure was £9,666 2s. 4d., inclusive of £6,983 2s. 10¾d. for material, and £1,498 18s. 9¼d for labour. At Sheerness the expenditure was £10,432 1s. 2d., inclusive of £7,393 16s. 7d. for material, and £1,388 19s. 8¾d. for labour. At Portsmouth the expenditure was £17,260 16s. 9d., inclusive of £12,179 11s. 3d. for material, and £2,411 6s. 1d. for labour. At Devonport the expenditure was £16,132 5s. 10¾d., inclusive of £11,665 10s. 1¾d. for material, and £2,592 4s. 7d. for labour. At Pembroke the expenditure was £66 14s. 2d., inclusive of £38 15s. 9d. for material, and £19 17s. 2d. for labour. Sails, like rope, admit of the perfect application of Sir

---

comes the question of the useless and excess stores. I do not alone refer to timber. In all the dockyards there are enormous storehouses in urgent need of gutting. In the one dockyard at Devonport, 400 men, aided by a powerful engine and fine machinery, spin and twist rope continually; as many probably are employed doing the same thing at Portsmouth and in each of the other dockyards. I am persuaded, from what I learned, that these men spin and twist more rope in one year than the navy requires in five years. So, also, in the manufacture of various other articles. Accumulation for war proceeds without reflection or control. Putting together all the guesses I could form in all the dockyards, from all the information I could gather, I arrived at the following sum total, which would be realised in the market by clearing out the excess accumulations in the dockyards:—

1. Timber excess...	£2,500,000
2. Cordage, sail, and mast excess	1,500,000
3. Returned stores excess	1,000,000
4. Dockyard stores excess	500,000
5. Contract stores excess	500,000

Inconvenient excess of Sundries................£6,000,000

—*The Dockyards and Shipyards of the Kingdom.*

George Lewis's tests, and the silly unreasoning accumulation of them in the dockyards calls loudly for restraint.

The Colour-lofts. Tenth in order stand the colour or bunting lofts. The charge for these for the year was £2,299 14s. 2¾d. At Woolwich the expenditure for material was £228 11s. 8d., and labour £95 1s. 7d.; total, £359 11s. 5d. At Chatham the expenditure for material was £164 11s. 1½d., and labour £156 15s. 3d.; total, £371 6s. 1¾d. At Sheerness the expenditure for material was £124 15s. 3d., and labour £122 10s. 6d.; total, £285 12s. 5¼d. At Portsmouth the expenditure for material was £297 18s. 9d., and labour £97 14s.; total, £467 8s. 6d. At Devonport the expenditure for material was £521 2s. 2¼d., and labour £172 2s. 3d.; total, £815 15s. 8¾d. That bunting might be advantageously supplied by contract is true, and that £2,299 4s. 2¾d. for flags annually is excessive, is also true.

The Rigging-houses. Eleventh in order stand the rigging-houses. The expenditure for these for the year was the large sum of £72,270 17s. 0¾d. At Woolwich the charge for material was £7,991 2s. 3d., and labour £1,578 15s. 4d.; total, £10,822 19s. 5d. At Chatham the charge for material was £11,838 16s. 0¾d., and labour £2,257 3s. 3¼d.; total, £15,556 8s. 10¼d. At Sheerness the charge for material was £12,077 17s. 7d., and labour £1,392 2s. 2d.; total, £16,131 5s. 10d. At Portsmouth the charge for material was £9,293 13s. 7d., and labour £2,280 0s. 11d.; total, £14,572 8s. 1d. At Devonport the charge for material was £11,245 7s., and labour £2,043 9s. 6d.; total, £15,587 14s. 10½d. Rigging-houses are necessaries at least for storage if not for preparation, but

£72,270 17s. 0¾d. of rigging annually is absurd. The quantity of rigging in store in the dockyards is so great that it is not likely to be ascertained until the store-houses have been pulled to pieces and cleared away, and further additions to the immensely overgrown stock cannot be justified even on the ground of timid prudence.

The Lead-mill.* Chatham has a monopoly of the dockyard lead manufacture. During the year the lead-mill turned out 21,852cwt. 1qr. 21lb., at an expenditure in all of £26,533 5s. 8¼d.

The Paint-mill. Chatham has also a monopoly of the dockyard paint manufacture. During the year the paint-mill worked up material to the extent of £15,373 16s. 11d., and £230 7s. 9d. was paid for labour: Inclusive of the extra charges the total expenditure was £17,385 15s. 9¼d.

The Metal-mills and Foundry.† Chatham has also the only metal-mills. These mills produce—in copper, sheets, and sheets for braziers, bolts for targets, bolt-staves, half-rounds, flats and square; in iron, bolt-staves, flats, angle, mast-hoop, squares; in metal, sheathing nails, castings, rings, deck-nails, bolt-nails, plate-nails; and brass sheets. The quantity of metal used was 51,035cwt. 3qr. 1lb., and the total expenditure for the year £264,567 9s. 9¼d.

---

* The Committee are of opinion that a machine for making lead pipes by pressure would be preferable to the present mode of drawing, as it would produce better pipe, in longer lengths, and at less cost.

† The Committee found, on their late inspection of these mills, that they were not worked to their full advantage in comparison with similar mills in private trade. At present the mills are worked with two gangs during the day, and only one gang at night, by which arrangement some of the furnaces are thrown out of work during the night, causing injury to the furnaces by cooling and re-heating, besides great waste of fuel.—Page 61, Report 1859 Committee.

The Cement-mill. Last of all, Chatham possesses the only dockyard cement-mill. For the year 160 tons and 180 tons of Harwich stone were "converted," at a cost of £376 2s. 8¾d.

The Blockmakers' Shops.* Sixteenth in order stand the blockmakers' shops. During the year 1860-61 the expenditure in these shops was the very large sum of £28,246 3s. 3½d. At Portsmouth there was an expenditure of £17,466 4s. 6d. for material, and £2,547 9s. 6d. for labour; total, £27,689 10s. 4d. At Devonport there was an expenditure of £156 14s. 2d. for material, and £278 13s. 10d. for labour; total, £736 12s. 11½d. At Deptford, Woolwich, Chatham, Sheerness, and Pembroke, "no separate accounts." Whether by the non-existence of separate accounts it is to be understood that five out of the seven dockyards keep their blockmakers' accounts together, or do not keep accounts at all, is, of course, for the Secretary of the Admiralty to say. So large an expenditure for an article that might be supplied when wanted by contract, is an anomaly that should be seen to.

The Trenail-houses. Seventeenth in order stand the trenail-houses. For the year the expenditure in these houses was £4,411 11s. 10¾d. At Deptford there was an expenditure of £9 2s. 6d. for material, and £19 9s. 3d. for labour; total, £32 12s. At Woolwich there was an expenditure of £544 2s. 1d. for material, and £86 0s. 10d. for labour; total, £704 17s. 6d. At Chatham there was an expenditure of £1,114 5s. 9¾d.

---

* The Committee consider that the men at present employed in the block-mills are not well classed or arranged with reference to their pay and the work on which they are employed.—Page 56, Report 1859 Committee.

DOCKYARD MANUFACTURES.

for material, and £286 13s. 2¾d. for labour; total, £1,648 14s. 8d. At Sheerness there was an expenditure of £14 10s. 1d. for material, and £34 19s. for labour; total, £56 9s. 0¾d. At Devonport there was an expenditure of £87 19s. 8½d. for material, and £30 12s. 4¾d. for labour; total, £144 6s. 6d. At Pembroke there was an expenditure of £1,401 8s. 10d. for material, and £188 12s. 11d. for labour; total, £1,824 12s. 2d.

The Oarmakers' Shops.* Eighteenth in order stand the oarmakers' shops. For the year the expenditure in these shops was £6,243 9s. 6½d. At Chatham there was an expenditure of £4,162 6s. 5½d. for material, and £563 7s. 2¼d. for labour; total, £5,455 7s. 4d. At Portsmouth there was an expenditure of £39 4s. 8d. for material, and £52 10s. for labour; total, £100 8s. 1d. At Devonport there was an expenditure of £391 0s. 3d. for material, and £211 7s. 4d. for labour; total, £687 4s. 1½d.

Caulkers and Pitch-heaters' Shops. Nineteenth in order stand the caulkers and pitch-heater's shops. For the year the expenditure in these shops was £11,379 12s. 4d. At Chatham there was an expenditure of £1,256 1s. for material, and £84 10s. 1½d. for labour. At Portsmouth there was an expenditure of £4,583 13s. 3d. for material, and £755 11s. 8d. for labour. At Devonport there was an expenditure of £2,222 18s. 11d. for material, and £175 17s. for labour. At Pembroke there was an expenditure of £1,075 14s. 11d. for material, and £93 19s. 3d. for labour.

---

* The oar-making machine at Chatham being capable of manufacturing sufficient oars to meet the wants of the whole service, no further machinery for this purpose is necessary; consequently the making of oars by hand should be discontinued generally. —Page 56, Report 1859 Committee.

The Turners' Shops. Twentieth in order stand the turners' shops. For the year the expenditure in these shops was £2,340 6s. 11½d. At Deptford there was an expenditure of £413 15s. 10d. for material, and £67 10s. 7d. for labour; total, £481 6s. 5d. At Woolwich there was an expenditure of £6 7s. 9d. for material, and £12 7s. 2d. for labour. At Chatham there was an expenditure of £262 16s. 6½d. for material, and £63 1s. 11¼d. for labour. At Sheerness there was an expenditure of £633 4s. 5¾d. for material, and £118 16s. 2¼d. for labour; total, £1,475 15s. 6d. At Devonport there was an expenditure of £171 3s. 8¼d. for material, and £101 17s. 4d. for labour; total, £2,647 14s. 2d.

Locksmiths', &c., Shops. Twenty-first in order stand the locksmiths', &c., shops. For the year the expenditure in these shops was £2,119 5s. 7d. At Woolwich there was an expenditure of £574 17s. for material, and £318 3s. 5¾d. for labour. At Sheerness there was an expenditure of £177 2s. 0¾d. for material, and £130 10s. 8d. for labour. At Portsmouth there was an expenditure of £221 14s. 1d. for material, and £231 9s. 4d. for labour. At Devonport there was an expenditure of £9 17s. 9½d. for material, and £84 13s. 7½d. for labour.

The Foundries. Twenty-second in order are the foundries. For the year the expenditure in the foundries was £5,862 4s. 9¾d. At Chatham there was an expenditure of £2,114 1s. 1¾d., inclusive of £953 5s. 4d. for material, and £885 11s. 2d. for labour. At Devonport there was an expenditure of £1,311 18s. 11d., inclusive of £880 9s. 11d. for material, and £252 9s. 4½d. for labour. At Pembroke (metal castings) there was an expenditure of £1,723 11s. 10d., inclusive of £1,266 0s. 9d. for material, and £207 8s. 2d. for labour; and at Pembroke (iron castings) there was an expenditure of £712 12s. 11d., inclusive of £221 5s. 2d. for material, and £316 13s. 1d. for labour.

Hosemakers' Shops. Twenty-third in order stand the hosemakers' shops. For the year the expenditure in these shops was £6,837 15s. 7½d. At Deptford there was an expenditure of £1,160 7s. 3d. for material, and £74 9s. 4d. for labour. At Woolwich there was an expenditure of £642 5s. 2¾d. for material, and £86 14s. 6¾d. for labour. At Chatham there was an expenditure of £573 16s. 0½d. for material, and £37 10s. 8¼d. for labour. At Sheerness there was an expenditure of £596 19s. 3¼d. for material, and £61 7s. 7d. for labour. At Portsmouth there was an expenditure of £1,156 18s. 5d. for material, and £105 16s. 6d. for labour. At Devonport there was an expenditure of £1,258 19s. for material, and £296 12s. 7d. for labour.

The Painters' Shops.* Twenty-fourth in order stand the painters' shops. For the year the expenditure in these shops was £5,170 3s. 7¾d. At Deptford there was an expenditure of £140 18s. for material, and £12 14s. 11d. for labour. At Woolwich there was an expenditure of £230 9s. 4d. for material, and £31 2s. 6d. for labour. At Chatham there was an expenditure of £51 10s. 10¼d. for material, and £24 17s. 7d. for labour. At Sheerness

---

* The present practice of employing labourers to do painting to so great an extent is expensive in material and productive of inferior work. The Committee could not but notice the small supervision over this class of men compared with the large amount in all other cases, and the variety of pay among the leading men of painters at the different yards.—Page 51, Report 1859 Committee.

there was an expenditure of £627 5s. 3¾d. for material, and £283 12s. 9¾d. for labour. At Portsmouth there was an expenditure of £1,077 9s. 8d. for material, and £523 2s. 2d. for labour. At Devonport there was an expenditure of £880 19s. 5½d. for material, and £409 0s. 0¼d. for labour. At Pembroke there was an expenditure of £316 13s. for material, and £3 6s. 5d. for labour.

The Condensor's Shop. There is only one condensor's shop, and that is at Portsmouth. For the year the expenditure was £15,831 18s. 8d. Among the manufactures were 2,000 metal screw bolts, 1,910 copper ditto, 1 galvanised iron bucket, 1 bath, 6 hammers, 4 coupling irons, 1 pair steel shears, 1 copper ventilator. The condensor's shop is evidently a job of some considerable proportions. Who ordered the bath? Who required the steel shears? Who fancied a copper ventilator? For what purpose was one galvanised iron bucket intended? It is worth repeating, the condensor's shop is an annual charge of £15,831 18s. 8d.

The Pump-house. Portsmouth also has a pump-house, and no other dockyard has one. The Portsmouth pump-house, in return for £840 10s. 0¾d., supplied 1,236 feet of hand pumps, and performed an inconsiderable amount of repairing.

The Fire-engine Shops. Devonport alone furnishes an account of its fire-engine shop, the other dockyards no doubt being ashamed to do so. The Devonport shop was a charge of £145 6s. 4d., and that sum yielded several leather cups, a number of leather washers, and three leather valves. Why Bermondsey was not applied to, and why the fire-engine shops of the other dockyards can be omitted from a statement, are of course for the Secretary of the Admiralty to explain.

The Smitheries.* Twenty-eighth in order stand the smitheries. For the year the expenditure in the smitheries was the large sum of £211,606 18s. 5¼d. At Deptford there was an expenditure of £14,511 3s. 4d., inclusive of £4,526 19s. 2d. for material, and £5,331 17s. 9d. for labour. At Woolwich there was an expenditure of £27,743 15s. 8d., inclusive of £14,191 1s. for material, and £9,101 16s. 1¼d. for labour. At Chatham there was an expenditure of £32,884 7s. 6¼d., inclusive of £12,887 10s. 0¼d. for material, and £14,062 9s. 6d. for labour. At Sheerness there was an expenditure of £18,167 16s. 2d., inclusive of £6,842 11s. 4d. for material, and £8,532 15s. 6¾d. for labour. At Portsmouth there was an expenditure of £55,063 4s. 4d., inclusive of £18,171 6s. 9d. for material, and £25,082 15s. 6d. for labour. At the Devonport South Smithery there was an expenditure of £34,580 16s. 3d., and at the Devonport North Smithery there was an expenditure of £14,955 1s. 8d. At Pembroke there was an expenditure of £13,700 13s. 6d.

The Steam-hammer Shops. Twenty-ninth and last in order stand the steam-hammer shops. For the year the expenditure in these shops was £50,537 9s. 11¾d. At Deptford there was an expenditure of £3,102 9s. 7d., inclusive of £1,296 9s. 3d. for material, and £384 for labour. At Woolwich there was an

---

* The Committee are of opinion that the smitheries generally are not in such a state of efficiency as the Admiralty have a right to expect. The Committee found that iron beams for ships are being manufactured at some of the yards from blooms at a great and unnecessary expense, as iron of a form capable of producing a stronger beam at a less cost is procurable under the contract.—Page 58, Report of the 1859 Committee.

expenditure of £7,073 15s. 4½d., inclusive of material £3,273 2s. 6d., and labour £1,901 1s. 6d. At Chatham there was an expenditure of £7,327 15s. 8¾d., inclusive of £5,320 7s. 7¾d. for material, and £1,317 2s. 8d. for labour. At Portsmouth there was an expenditure of £13,302 16s. 1d., inclusive of £6,097 3s. 3d. for material, and £2,210 6s. 1d. for labour. At Devonport there was an expenditure of £11,089 12s. 7d. At Pembroke there was an expenditure of £4,243 9s. 9d.

Conclusions. Such is the long and tedious catalogue of the manufactures of the dockyards. By their own showing the Admiralty carry on no fewer than twenty-nine trades, all of which no doubt were at some time or other ancillary to shipbuilding, but most of which have not now-a-days anything to do with shipbuilding. Seeing that in private enterprise the more concentrated supervision and effort are, in other words, the fewer the irons in the fire the better, this multiplicity of occupations is a mistake. It is incompatible with a good working system. Nay, it is worse, for it represents the dockyards in the ridiculous light of national repositories or museums, where all the occupations that have been ever followed are studiously conserved. When, therefore, the reform axe comes to be laid to the dockyard tree, all the superfluous occupations must be weeded out. The plant must be sold, the brick and mortar walls demolished, and the superseded men discharged; no doubt in the existing and most improper manner. In the vacant spaces left by the removal of useless and costly buildings the Junior Lord of the Admiralty may excavate docks or basins, and if these are not wanted let him offer freehold land for sale, or make gifts of land for widening the streets and otherwise improving our very filthy and very ugly dockyard towns.

Chapter IV.

DOCKYARD SHIPBUILDING.

Scientific uncertainty.* Shipbuilding, although based on scientific principles, has so insecure and indefinite a hold on them, that it may be regarded as occupying one of the least satisfactory positions among the arts. In a certain sense there is still contention as to the purely elementary principles of shipbuilding, so that it might be said to have no claim to be numbered among the arts at

---

* No branch of art is so overlaid with traditional prejudices as shipbuilding; and it is to the operation of these prejudices through several generations, to the influence of absurd and restrictive laws, to the continual and systematic impediments thrown in the way of improvement by a bigoted and obstructive Navy Board, that we must ascribe the indifferent progress made in our royal and mercantile navies for more than a century after the time of James II., from whose reign, curiously enough, we can trace a direct line of Surveyors of the Navy. While the French encouraged improvements from every quarter, the Navy Board officially censured ingenious officers who, through honest zeal, pressed their claims too warmly. Instances of this are on record.—Pages 100 and 101, *Memoirs of Rear-Admiral Sir William Symonds, Surveyor of the Navy from 1832 to 1847*.

In Committee on the Navy Estimates, 15th April, 1836, Mr. G. F. Young objected to the vote for stores, because he believed there was no responsibility that they would be properly applied. He complained, too, of the lavish expenditure of stores occasioned by the experiments in naval architecture now being made by Captain Symonds —not then Rear-Admiral. He had a great respect for that officer, but he thought that gentlemen brought up to the business of shipbuilding were more likely to build ships well than a naval officer with whom every ship he constructed was an experiment. Captain Symonds claimed to have discovered a new principle of shipbuilding; for his part he did not believe in its efficacy; but all the old ships would be pulled to pieces to build new ships on Captain Symonds' plan.—Pages 182 and 183 of the same Memoirs.

all; and in another sense shipbuilding may be considered as having attained a development and perfection to which no other art can aspire. Look, for example, at the *Great Eastern*—a perfect prodigy in its way. Where is there another monument of genius in the wide world that may fitly be compared with it?* The Thames Tunnel, the Menai Bridge, our railways, mines, foundries, and mills are prosaic and unworthy by its side; for, superadded to the wonders of calculation, combination, and contrivance, is the purpose to which the great ship is applied:—

> "Her march is o'er the mountain wave,
> Her home is on the deep."

And in not a much less degree the same is to be said of the ocean steamers built for the Cunard, the Peninsular and Oriental, the West India Mail Company, and many others. Look at the lines of those noble vessels, observe their entrance and departure from a wave, watch their motions in troubled water, and say where they are imperfect! Criticism is dumb. The world has never seen such ships, and we cannot bring ourselves honestly to think that the world will ever see much better. Is not this, then, the perfection of art? Think as we may of troubled water, steadiness, and buoyancy, if we cannot suggest improvement, and if to us the forms of these superb vessels seem final, where can science possibly be in error? Still these questions may be answered. Proud as we well may be of the various types of merchant ships at the present time afloat, it is known to all who think of such things that too often science in a strict sense has no considerable share in these results. Our

---

* In France I found that Mr. Scott Russell is held in very high estimation, and that the *Great Eastern* there will always be deemed a marvel.

best ships are in the main the creation of instinct rather than of rule, as regards their better qualities. Of shipbuilding, as it exists at present, it is not less true than too often it is of writing, that he who slavishly adheres to rule displays pedantry at every turn, and in all likelihood produces that which neither meets his own expectations nor the desires of others. Such a state of things is unsatisfactory; and the best of all proofs that it in reality exists is to be found in the fact that four years ago an Institute of Naval Architects was founded, which as yet has not succeeded in grappling with, much less overcoming, the acknowledged evil.

<small>The problem of Shipbuilding.</small> The problem of shipbuilding is a plain one, and it may perhaps be best stated in connection with the construction of ironclads. Which is the best structure for steadiness, speed, and carrying capacity, on a small or great draught of water?* But the problem may be stated without reference to ironclads, inasmuch as it was present to men's minds before ironclads were thought of. It then took the form of speed under sail or steam, and capacity for coal or cargo, or both together. Manifestly, were the problem solved, there would be no further occasion for rule-of-thumb shipbuilding, and in the case of ironclad construction there would be all the certainty that there now is in building bridges or steam-engines. Any one might then build a ship, just as any tailor makes a pair of trousers. In a word, all doubt and

---

* The recent very interesting and instructive debate at the Institute of Naval Architects left in doubt whether Mr. E. J. Reed's *Enterprise*, a ship with a belt of armour round and round, and a square armoured fighting-box on the *Warrior* plan, with broadside ports in the centre, was a preferable ship to Captain Coles's *Naughty Child*, armoured round and round and with its armament within revolving turrets.

mystery would be removed, and ships provided to answer in perfection their intended purposes.

<small>The method of solution followed.</small> The method of solution followed with unvarying uniformity and astonishing complacency has only to be hinted at to be at once familiar to those even superficially acquainted with the philosophy of science. The higher mathematics is the first and almost the sole study of those who would discourse acceptably at the highly interesting meetings of the Society of Naval Architects. Those with only an imperfect knowledge of mathematics will do well to keep their seats, because, if they presume to rise, they will convince no one, and there is some danger of their being listened to by a visibly impatient audience, if not positively by empty benches. The members of the Society of Naval Architects adore mathematics and they adore nothing else. Mathematics, therefore, is their method of solution. He who would shed light on the great problem of shipbuilding must soar high above the common herd in the regions of abstraction. Is not this the philosophy of the Greek schools, which failed so lamentably? When Thales was asked, "What is the *greatest* thing?" he replied, "*Place;* for all other things are *in* the world, but the world is *in* it." Aristotle, again, when inquiring what *Place* is, observes, "If about a body there be another body including it, it is in *place;* and if not, not." Whether this nonsense was or was not intelligible to those to whom it was addressed is unimportant, the point being that it accomplished nothing. It was hitching up a dead horse. The ideas of the Greeks, as Whewell expresses it, were not distinct and appropriate to the facts. Facts were not disregarded, but instead of being employed as bricks and

mortar in the hands of a bricklayer, the Greek philosophers sat on their stools chewing the facts in much the same fashion as heifers chew their grass. If, when Herodotus, in attempting to account for the floods of the Nile, got hold of the notion of the attraction of the sun, he had extended his observation, the discovery would have been made that the hypothesis was unsound. So with his other conjectures, and, following the course of conjecture and verification, he would no doubt have arrived eventually at some satisfactory explanation. Now, mathematics is a proper study for shipbuilders as it is for most classes, although some of the best ships afloat have been built by those who know nothing and care nothing for mathematics. But mathematics is an inappropriate idea to the undetermined facts of shipbuilding. A lecturer on astronomy might as well confine his observations to the telescope, or a lecturer on anatomy to the character of the subject stretched on the dissecting table, and which probably was brought from Newgate. Mathematics is to the shipbuilder less than the yard measure to the draper; and certainly a draper who, instead of serving customers, spent his time writing and reading papers on the yard measure, would be thought a silly fellow. What shipbuilding really stands in need of is complete and perpetual emancipation from mathematics. To accomplish this, that which Herodotus needed to enable him to account for the floods of the Nile—namely, observation—is alone required. To nothing else but learned trifling is it the case that the best of all treatises on floating bodies yet published in in this year of grace 1863, is that of Archimedes.

<small>What might be done.</small> What might be done to give certainty to shipbuilding has been suggested by an ex-

cellent although unassuming treatise.* Mr. Bland, in writing on the form of ships and boats, introduces us to experiments, any one of which is practically of more value than a cartload of papers in the muddled and generally unknown tongue of the higher mathematics. Experiment, or its observation, is the royal road to excellence and simplicity in shipbuilding, and if it is objected that Mr. Bland's experiments were on smooth water, then let this be remedied by experiments on troubled water. Such experiments are no doubt as much within the means of the Institute of Naval Architects, as they ought to be within its scope. Mr. Froude and others, accomplished mathematicians no doubt, superseding the necessity of practical experiment when water, wood, and iron are the simple elements, is as monstrous as a mountebank telling fortunes.† But for their extraordinary pretensions, it is likely that men of a more philosophical turn would, even in the few years that have elapsed since what may be called the revival of science in shipbuilding, have struck out in that true direction in which

---

* *Hints on the Principles which should regulate the Form of Ships and Boats, derived from original experiments.* By W. Bland, Esq. John Weale, 59, High Holborn. Price 1s.

† Three " native artists," who had gained the highest prizes in mathematics and written the best essays, were sent to Chatham to meet, and examine the building of ships and propose the proper lines for each class. They handed in (the year before this, 1848) a very elaborate statement " giving reasons why *every ship had faults*, and why *perfection had not been obtained.*" These gentlemen were now to build a ship themselves " in order to show whether they could not produce *the best ship in the world.*" There is really a touch of humour in this declaration which makes one smile, especially when the rather exalted pretensions of the Scientific Committee are measured by their actual performances. The *Espiègle*, their most finished piece of handywork, stood third in the trial of experimented brigs at the end of this year; and the *Thetis*, which they afterwards built, was, considering all things, a most egregious failure.—Page 315, *Memoirs of Sir William Symonds.*

Mr. Edye, in his evidence before the Committee on Naval Estimates, 1847-8, says of the same parties, "In two or three cases they may have given a theoretical report on drawings; but I am not aware of any benefit that has yet been derived or any *science that has been displayed by them.*"

every step is progress, and a permanent addition to the stock of human knowledge. Such philosophical inquiries might by this time have given us all the possible calculations of all the possible forms of ships, in a ready reckoner to which every shipbuilder might have referred when necessary. Nor need they have stopped there. To the ready reckoner there might have been appended the certain behaviour of all possible forms of ships under all possible circumstances, both in smooth and troubled water. But such men—and among the intelligent shipbuilders of the kingdom there must be many such—have been restrained, nay, put down, by the Greek sages who impudently have elbowed their way among them. This is why the anomaly exists of ships of all classes being built half-way by science and half-way by instinct. Shipbuilders use science just as a prudent sea-side bather uses deep water, that is,—to the extent he safely can; and then they have recourse to the rough but ready experience of the performances of other ships, which they have casually, and perhaps imperfectly, gathered from conversations or reports. That this should be so, that the experience of all men, and that which all men have still to learn experimentally, as regards certain forms of ships, should to this day remain unsystematised and unprovided, can be regarded as nothing less than strange. Until it has been supplied, the less we hear of naval architecture and of schools of naval architecture, the more will it be to our credit.

The Dockyards in the main to blame. That the dockyards are in the main to blame will naturally occur to many. The anticipation is too true. Until the other day the master shipwrights of the dockyards were the great authorities on shipbuilding. What these gentlemen considered proper was orthodox, and what they had under consideration, or what they had condemned, was heterodox. For any private shipbuilding firm to be deemed worthy to supply a broomstick to the navy, it had to be in full dockyard communion. The master shipwrights were the bishops and apostles of the profession, furnishing the elementary and advanced text-books, and consecrating private shipyards for public jobs. Their day has now gone by. Even the Admiralty are alive to the professional unworthiness of their own long-tried and extraordinarily overrated advisers in all that relates to shipbuilding matters. The private shipyards are now admittedly the great depositories of shipbuilding knowledge, and Mr. E. J. Reed is the first draft on the healthy and abundant shipyard stock.* As each master shipwright, and assistant master shipwright, falls back in mortification and disgust on his unearned and iniquitous superannuation of £300 to £600 per annum, let his place be filled in the same manner from the shipyards, keeping back the hungry dockyard tide of incompetent servility and superannuation that insolently claims each opening as its own. As long as dockyard genius was in the ascendant and had everything its own way, shipbuilding science made no headway. Independent inquiry and speculation were under a rigorous ban. Theory within the dockyards was proscribed as sternly as without. The Surveyor of the Navy, and the dockyard underlings dancing attendance on him, while complacently asserting their great superiority and pre-eminence, claimed to be merely practical. Long service afloat had polished and matured the Surveyor's judgment on the

---

* Mr. Reed is no doubt of the dockyard stock, but had he remained in the dockyards he never would have filled the high position that he does. His training has been among the shipyards and in his own chamber.

qualities of seagoing ships, and what the Admiralty had to sanction, and the constructors of the navy to carry out, was these sapient views. The system was the very antithesis of that which the Institute of Naval Architects is doing its best at the present time to perpetuate. And it was a system very nearly as inimical to real progress. It was substantially this. An old sailor thought that ships should be long or short, broad or narrow, deep or shallow, at the ends square or tapering, at the sides straight, full or cut away, and on the bottom flat slightly rounded, middling rounded, fully rounded, slightly wedge-shaped, middling wedge-shaped, largely wedge-shaped, &c. He was usually a man of one idea, or at most of two ideas, and the one or the two ideas were practical,—that is, the one idea might be full bows or tapering stems, and the two ideas one or other of these previous qualities, while the other might be wedge bottoms or flat bottoms. So ships were built to embody the one or the two ideas as it happened, and the ships being years in construction, and afterwards years in being tried, the scientific value of the one or of the two ideas was really nothing. And yet this wretchedly unphilosophical system was lauded in Parliament and out of Parliament, and its wretched authors had honours, pay, and pensions heaped upon them. No wonder that inquiry was stifled, and that there has been a revulsion of feeling even at the Admiralty. But it is discreditable to the Admiralty, and another conclusive proof of unfitness for the duties to be performed, that the narrow-minded practical men have been thrown overboard, only to embrace impracticable and misleading theorists. Making a show of caution and deliberation, the happy mean is, however, passed, and one extreme is exchanged for another.

Effects of such a state of things. There are various palpable effects of such a state of things, some of which may be referred to in a few words with profit. As long as the generality of men in the dockyards and out of them recognised in Surveyors of the Navy and in master shipwrights men of surpassing genius, it really was not worth the while of any one to think of dockyard or ordinary shipbuilding matters. The practical motto was, "Let well alone." Everything was thought of unofficially, just as everything was thought of officially. Therefore it will be found that in very plain matters, such as all are presumed to know accurately about in other things, public opinion is still vague and inert as regards shipbuilding. As regards shipbuilding, the off-hand logic of men who might understand the subject properly with scarcely any effort, is slipshod and inconclusive. The ground they are treading on is still felt to be sacred. Take an instance. Up to this time men of all classes who speak and write of the dockyards are in the habit of citing France and America, as if England ought to do likewise on all occasions. France, it may be said, is building wooden-bottomed ships, therefore England ought to build wooden-bottomed ships. Or the Americans are building enormous ironclads, therefore England ought to build enormous ironclads. Here it is assumed that what France and America are doing is right. But, very oddly, the truth is that France and America are at the same time reasoning about ourselves in the same manner. America has constructed, and is still constructing, ships of war of preposterous tonnage, simply because England is constructing ships of war of preposterous tonnage; and France adheres to a combination of wood and iron because England adheres to it. Opinion, as regards these things, is thus moving in a circle. Then, again, it seems generally to be overlooked that England is an iron-producing and iron-manufacturing country, and that America and France are neither. America, therefore, has an interest in using wood, although its use may be to some extent a disadvantage; and who knows but that France, from the nature of its American commerce, exporting silks, wines, and the like, and standing in little or no need of flour, butter, or provisions in exchange, may not also have a similar motive to the use of wood, although under different circumstances it might have given an unqualified adhesion to iron? It may thus be wisdom on the part of timber-importing France and of timber-growing America to continue the construction of timber or of timber-bottomed ships, while it may be equally the part of iron-producing, iron-manufacturing England to give up timber shipbuilding altogether. Take another instance. Why Russia builds ships of war in the Imperial dockyards arises solely from the fact that there are no great private shipbuilding yards in the country. Why France builds its ships of war in the dockyards of the Empire is owing to the inadequacy of the great private shipbuilding yards; and why, on the other hand, America builds its ships of war in the private shipbuilding yards, is because the public dockyards are inadequate. Further, why Italy builds its ships of war in France, England, and America, is to be accounted for by the want both of dockyards and shipyards. Still, let any one say a word against our dockyards, and he is at once met by the statement that all the great naval Powers possess great dockyards, which is untrue; or by the statement that without dockyards there can be no naval power, which the case of Italy—soon to be a most formidable naval Power—at once satisfactorily disproves.

But it is needless to pursue the subject. People have yet to learn to judge of all that appertains to shipbuilding, as they have learned to judge of almost every other thing. When they have done so they will think only of the merits of each separate case, perfectly regardless what France, America, Russia, or Italy are doing. Does a City shopkeeper care about the Paris or the New York modes of making money? No. He wisely adapts himself to the circumstances in which he is. Why, then, should the British navy be provided for in any other manner?

<small>Shipbuilding under the practical *régime*.</small> Under the practical *régime* dockyard shipbuilding never flourished. There were abortions of all kinds turned out, ships that would neither steam nor sail, and as some approached completion it was not uncommon to discover that there was a screw loose somewhere; in other words, that the ship would not carry the intended weights, or positively would not swim. Nor are errors of this kind at all old dated, for almost one of the last official acts of one of the last highly honoured great dockyard men was to propose to the present Board of Admiralty the construction of an ironclad frigate possessing no greater floating properties than a cannon-ball. But to return. When the discovery happened to be made that a ship on the stocks would not swim, two courses presented themselves to the professional advisers of the Admiralty. The first was to cut the ship in two and increase the length sufficiently; the second to allow the unlaunched ship to remain and rot, although rotting is a very tedious process, at least when prayed for. But it did not always follow that a spoiled heavy frigate would make a good line-of-battle ship by twenty or thirty feet being added

to the length: sometimes such frigates would only cut up into bad corvettes. In this last case these bad corvettes literally cost the country, if the truth were told, or could be told, their weight in gold. In the first place, they might have represented the whole palpable produce of the labour of the dockyard, and within the walls there would in no case be fewer than 2,000 men, and possibly there might be rather more than twice as many. Fancy, then, the spoiled ship in hand a single year, and the first cost is a heavy one. Then comes the conversion; and that, as things went and still go in the dockyards, might once more be the one visible effort of all hands during as long a period as before. Thus the construction of one such ship represented the use and waste of a perfect mine of wealth; the amount probably exceeding the yearly value of the produce of some rural parishes. The introduction of steam, and afterwards the supersession of the paddle by the screw, were great occasions for the practicals. Both were godsends in their way, inasmuch as they were changes from the old routine, and in addition to filling the hands of all with work, were exceedingly beneficial to the shopkeepers, landlords, and others of the dockyard towns. Admiral, Commodore, and Captain-Superintendents felt as eagles do when age is renewed to them. Surveyors of the Navy were equally elated. The introduction of machinery of course led to one and all of those mishaps that may readily be surmised. Between the heads of departments in Whitehall and Somerset House, and the heads of departments in the dockyards, there was the same sweet harmony as is got out of a fiddle without the first and second strings, or out of a hurdy-gurdy on a rainy day. The engines were occasionally too large or too small for the boilers, and not unfrequently too

K

large or too small for the ships for which they were designed. The Select Committee appointed to inquire into the expenditure of the Army, Navy, and Ordnance, indeed discovered that so powerful in some cases was the machinery ordered for weak and rotten ships, that had it been fitted the ships would have been torn to pieces, while in other cases so little motive power was ordered for some ships that propulsion against wind or tide would have been impossible. These blunders apply, however, more particularly to the first conversion to paddle ships. The introduction of the screw extracted the same discord from the official hurdy-gurdy. The sterns of ships had now to be torn out, and doing this was generally much the same thing as putting new legs to a pair of trousers, or new tails to a dress coat that has seen service. The old and new work could not be got to stick well together, and among nautical men it was long expected that the sterns of some of the so-called crack line-of-battle ships would fall away. The whole "guts" of the ships had besides to be torn out for the passage of the shaft, and when that was done the wrong engines, the wrong boilers, or unsuitable engines and boilers, were almost sure to be sent for fitting.

<small>Discovery of the 1859 Committee.*</small> The 1859 Committee made a remarkable and still unaccounted-for discovery as regards the relative price of ships built in the different dockyards during the practical *régime*. It was sub-

---

* The *Shannon*, a frigate of 51 guns and 2,651 tons, built at Portsmouth, cost £14,033 for shipwrights' labour, or at the rate of £5 5s. 10d. per ton, while the *Chesapeake*, also a frigate of 51 guns, but of 2,355 tons, built at Chatham, cost £9,372, or at the rate of £3 19s. 7d. per ton. The *Mersey*, 40 guns, built at Chatham and fitted at Portsmouth, cost for shipwrights' work £14,842, or £3 19s. 7½d. per ton, while the *Orlando*, of 50 guns, built at Pembroke and fitted at Devonport, cost for shipwrights' work £19,505, or £5 4s. 8d. per ton, though both ships are of the same tonnage, form, and dimensions.—Pages 18 and 19, Report 1859 Committee.

stantially this. In all the dockyards there were the same raw material, the same timber, tools, smiths' fires, &c., and the same classes of skilled and unskilled workmen engaged on ships, as like each other as peas or sheep. These skilled and unskilled workmen in addition received the same wages for the same tasks and jobs. But it was found that notwithstanding the identity of the conditions, the same ships cost twice as much in some of the dockyards as in the others. How this happened no one knew, and no one yet knows. If one baker charges sixpence for a four-pound loaf, and another charges two or three pence more, it is intelligible that quality, a credit business, or a higher rate of profit explains the difference. But what are we to think of two servants going at the same time into the same market, and the one giving us salmon at a shilling a pound, while the other charges twice that price? We would be disposed to put the severe construction on the case of the dear salmon—robbery. But the defaulting dockyards meet us with chapter and verse for all their outlays. So also have defaulters outside the dockyards until found out. The fraudulent clerk, secretary, manager, broker, banker, has always put a good face on things and made them pleasant.

Shipbuilding under the theorists.* With the practical *régime* let us now contrast that of the theorists. At the

---

* During the last session of Parliament a circular was distributed among members, by a mechanic, from which the following is an extract:—

"By the employment of wooden shipwrights to work on the iron war-ship *Achilles*, a great deal of money is uselessly squandered. I contend that the work can be more efficiently performed for one-third less than it now costs. Here is an instance of what I assert:—Where the wooden shipbuilders have been cutting holes for coal-shoots in the deck-plates, the same being three-eighths of an inch in thickness, size of hole 18½ inches in diameter, the time taken by them to cut one of these holes *was six days and a half*, when it would be done by men acquainted with iron shipbuilding *in five or six*

moment the Admiralty sweat and strain over the iron carcase of the *Achilles*. They are showing the public what the dockyards can do when they care to do anything. But the trick is too stale to deceive any one. A display of energy is inopportune when the fate both of the dockyards and the Admiralty trembles in the balance. It is a kind of death-bed conversion, which no one will regard as of much account. The effort is that of a shaky firm seeking an exchange of cheques, an exchange of signatures, or the purchase of anything on which money can be

---

hours at most. It is the same with the wooden shipwrights putting in rivets; it being well known that $41\frac{3}{4}$ rivets have cost £3 3s. 6d., when for 6s. or 7s. they could be well put in by practical iron-men. Now, as there will be a great number of rivets in the *Achilles*, *if the officials will persist in carrying the works on in the manner they do at present, there will be as much money squandered, besides material spoiled, as would build two such ships as the Achilles.*"

In reference to the manner in which the armour-plated ship *Caledonia*, is now building at Woolwich, the mechanic adds that an iron shipbuilder says—

"Her iron weather-deck is constructed of $\frac{1}{4}$-inch plates in thickness, riveted to iron girders, but owing to the unskilled manner the work is done by incompetent workmen, the plates not being properly levelled and riveted, the iron deck, when completed, presented one mass of buckles; so much so, that when they came to lay the wooden deck on the iron one, after shoring and wedging, they were compelled to reduce the 4-inch planks to $2\frac{1}{2}$, in various places, before they could lay them down. For all this blundering the real iron-men are blamed, when perhaps not one is employed, the woodenmen, because they happen to be at work on the vessel, being styled iron shipwrights, which is not the case,—and the public mind should be disabused of this."

On the other side, the *Morning Herald* and the *Standard* report of the launch of the *Delhi* from the Blackwall Yard of Messrs. Money Wigram and Sons, on the 10th of September, 1863, may be given:—

"In the construction there is no peculiarity beyond the somewhat important one that the iron *Delhi* has been built by wooden shipwrights, and that hereafter the firm of Messrs. Money Wigram and Sons are to be counted among the iron shipbuilders of the country. Messrs. Wigram's wooden shipwrights, appreciating the change that had taken place in their occupation, waited on the firm, and asked to be allowed to try their hands in the place of the boiler-makers, who are the usual builders of iron ships. The request was considerately complied with, first in the construction of a beacon for the Trinity Corporation, next in the construction of the *Diligente* for Brazils, an iron vessel for the Ganges, and the fine Dover passage-boats *Breeze* and *Wave* for the Chatham and Dover Railway; and, last of all, in the construction of the *Delhi*. With the result, owners, surveyors, builders, and workmen are all satisfied. The *Delhi* is pronounced a faultless piece of workmanship, and yet the work of wooden shipwrights."

raised through the friendly offices of the trade pawn-brokers, whose place of call should be a pillory in Threadneedle-street. The Duke of Somerset, Lord Clarence Paget, Admiral Robinson, and the others, must smile occasionally at their own hardihood. Rigging the market for the sale of Confederate cotton bonds is nothing to it. The *Morning Herald* reporter who supplied, for the second edition, an account of the execution of a notorious fellow who was favoured with a reprieve after his neck was in the fatal noose, was not a whit smarter. In the hurry of building the *Achilles* it is not surprising that one of the sides is said to be six inches out of truth. Nor is it to be wondered at if the workmanship compares as favourably with that of the great iron shipyards as the tin-kettle workmanship of Barnes Common compares with that of Lower East Smithfield. Providence, it is to be hoped, will, for the credit of the Admiralty, suspend those usually inexorable laws of nature which damn badly built and badly fastened ships, whether of wood or iron, when on the strand, or subjected to the rough usage of a whole gale of wind. Why the construction of the *Achilles* has provoked so little criticism arises from the building taking place in a dock barely large enough to hold the ship. Few care to undertake the descent to the bottom of the dock, and of that few not many are disposed to pursue knowledge under circumstances so embarrassing and filthy. Going down in a diving-bell to the foundations of Blackfriars Bridge is on the whole a more inviting undertaking than going down to the lower part of the *Achilles* among the blocks, props, smiths' fires, ashes, and other things below. It is, therefore, pretty clear why very little has been said of the workmanship of the *Achilles*. But it has been examined, and by practical men, who from disinterested motives pronounce it alike discreditable to the dockyards and the country. On the bottom, the unseen bottom of the *Achilles*, wooden shipwrights have been transformed into iron shipwrights. They knew nothing of iron work. They might as well have begun making shoes, making brick, or mixing mortar. And they laboured also under the trying disadvantage of having few learners, and these learners not being worth their salt as workmen. For some time before the iron *Achilles* was fairly taken in hand by the wooden shipwrights, gangs of them, as already stated, were sent for a week's, a fortnight's, or a month's instruction in the Thames Iron Shipbuilding Works, Orchard-street, Blackwall. This short apprenticeship in an entirely new branch of trade finished the training of the dockyard iron shipwrights, who are and were the instructors of their less favoured fellow-workmen. Such is the simple truth. By such hands has the iron *Achilles* been put together. By such hands two more iron ships are to be begun and carried on. True, they may now be thought fair workmen, but they certainly have had no claim to that title on the *Achilles*. The *Achilles* is as near an approach in workmanship to the other iron ships of the fleet as the letters gracing the doors and walls in an ordinary paint shop—the result of apprentices trying their hands—are an approach to the finished and faultless City signboards. But it may be that the Admiralty deny this, and indignantly scout the idea of the *Achilles* being six inches out of truth. Then the answer is, that the ship ought to be allowed to speak for itself.*

---

* A number of workmen are now busily employed at Chatham Dockyard in cutting away the solid granite sides at the entrance to No. 2 dock in which the iron ship *Achilles* is under construction, in readiness for that vessel being floated out. As much as 2 feet 8 inches, or 16 inches on each side, will have to be cut away from

Serious defects in the contract ironclads. In the construction of the *Valiant* class of ironclads there is one Admiralty defect that Mr. Scott Russell has condemned with great force and truth. There is a belt of armour round the gun deck of these ships fore and aft, and two more breadths of the belt extend along the whole length of the ship between perpendiculars lower down than the gun deck. The consequence must be that a shell lodged forward or aft of the two extra breadths of armour will blow those above in the protected gun deck into the air. At the expense of relieving the extremities of a great load, but in a downright stupid manner, a complete trap has been contrived, which in action promises to be destructive to our own sailors and spread dismay among the whole ship's crew. Another defect, and it is one that applies to all our ironclads, is insufficient ventilation. An ordinarily tall man cannot walk on the gun deck without stooping, and yet those decks are thronged with guns of large calibre. The moment, therefore, these ships go into action the crews will be oppressed with smoke, and it is conceivable that after a few rounds the firing must cease until the smoke clears away. Nay, it is impossible that firing can proceed for any length of time, so that, other things being equal, our ironclads in action will fall an easy prey to an enterprising enemy who merely enjoys more room and air. But this is only one form and evil of the deficient ventilation. In these ships no improvement whatever has been attempted in the interior fittings so as to guard against diseases of a contagious type. As in the old-fashioned man-of-war, so in the new, all on board the

---

the entire length of the solid wall on both sides the dock in order to admit of the *Achilles* being floated out without touching the sides.—*Times*' "Naval and Military Intelligence," 27th August, 1863.

ship breathe the same atmosphere, whether that is pure or impure, wholesome or diseased. Smallpox, yellow fever, &c., may, therefore, spread at will throughout our ironclads, although it must be an easy matter to devise means of isolation—means that would carry off impure air, and supply a healthy separate atmosphere at all times to sick and well. In an excellent pamphlet published some short time ago* the facility of fitting ships in such a way was pointed out, and it may be well to state that Mr. Lungley has for years been extensively employed adapting merchant ships for the transport of troops, emigrants, convicts, &c., in the usual manner. On such a subject he, therefore, speaks with authority, and it is surprising that his suggestions have not been acted on by the Admiralty. But Mr. Lungley's suggestions do not stop with thorough ventilation; his mode of ventilation is also calculated to keep a ship afloat when otherwise it would sink. To each deck he not only furnishes a watertight trunkway, but he does so to each separate compartment of each deck, so that a ship is formed of many parts perfectly detached from each other; and if to each compartment there are two such watertight trunkways, it is manifest that a constant and separate current of pure air must always be maintained. That not one improvement of any practical value should have been adopted in our ironclad fleet cannot fail to excite indignation and surprise.

The want of small Ships. The want of small ironclads is another sin. With ample warning and sufficient time to produce a brood of sloops of small tonnage and light draught of water, no more than three are yet in

---

* *Unsinkable Shipbuilding in its Applications to Iron and Wooden Ships of War, and Troop, Transport, Emigrant, and Merchant Vessels.* By Charles Lungley, Deptford Green, Dockyard, 1861.

hand—namely, the *Enterprise* and *Favourite* at Deptford, and the *Research* at Pembroke. The consequence must be, were war to arise, that the same state of things would be repeated as when Sir Charles Napier went to the Baltic. Against no Power whatever could we operate successfully on the coast with our *Minotaurs*, our *Valiants*, or our *Warriors*, and were war at any time to be forced on us by America, we would be lamentably weak. For the American lakes and rivers, as will appear hereafter, a smaller class of ships than even the *Enterprise* or the *Research* would be required, and should they have to be waited for after war were declared, Canada would inevitably be overrun, if not absolutely wrested from the Crown. And, strange to say, ten ships of the *Enterprise* class may be constructed for one *Minotaur*. Expense is, therefore, no excuse, and the omission or neglect can be explained only by the Admiralty and their advisers being strangers to common sense.

<small>The case of the *Dalhousie*.*</small> Shortcoming is crowned by the case of the *Dalhousie*, the ship constructed to commemorate the services of an Indian statesman during a trying period. The frigate *Dalhousie* was ordered in the usual manner; the Indian dockyard authorities set to work converting timber, setting up the frame and planking it, while Messrs. Maudslay, Sons, and Field, the Lambeth engineers, took in hand the engines and the boilers. In course of time the engines, boilers, &c., were received in store at Woolwich, and sent to India on board a large merchant ship, which had almost to be taken to pieces before the engines, boilers, &c., could be put on board. A few months'

---

* This case is given from the pamphlet on "the Dockyards and Shipyards of the Kingdom," page 16.

sailing carried the transport to its destination. The engines, boilers, &c., were landed, and the dockyard authorities looked forward to having the *Dalhousie* soon afloat and on the trial in the Indian Stokes Bay. They were doomed to disappointment. The Indian Office wrote to the Admiralty, and the Admiralty wrote to the Indian dockyard authorities to pull the ship to pieces, and send the pieces, machinery and all, to Woolwich. So the *Dalhousie* arrived at Woolwich last June (1862), and may be seen any day in timber and machinery stacks in Woolwich Dockyard.

<small>The system practically.</small> The writer is indebted to a dockyard officer for the following statement of the dockyard shipbuilding system as it is practically:—" The designs for building are all got out by the Controller's department at the Admiralty, and the form is the mere whim or fancy of the Controller, &c. The dimensions, proportions, and volume of displacement are arranged by the constructor and his assistants to suit the requirements of the vessel. Thus the proper function of the constructor is merely to get afloat the whims of the Controller, &c. The constructor and his assistants are generally men well qualified for their station, although there have been constructors who knew nothing at all of the scientific principles of shipbuilding, and made sad havoc with material. When the design by drafts and specifications is ready, it is signed by the constructor and Controller, and sent by the Admiralty to one of the yards, with an order for the vessel to be commenced. The Superintendent delivers the order, with the drafts and specifications, to the master shipwright, with whom they remain till the vessel is completed, when they are sometimes sent back to the Admiralty, and

sometimes not. The only thing then done by the master shipwright is to send the drafts, with a copy of part of the specifications, called the 'scheme of scantling,' to the single station day-pay shipwright, called chief draftsman of the mould-loft, who is under the direct orders of the master shipwright, but subject to the supervision of the foreman of shipwrights, who has charge of the new work. The chief draftsman and his assistants 'lay off,' or draw all the lines on the mould-loft floor, to the full size, and get all the section lines and measurements necessary for making the moulds, to cut the frames and erect them in accordance with the design. The moulds are now made by the joiners, who send them to the converters, and these draw timber by a demand note from the storekeeper, under the authority of a timber inspector. The frame timbers are then cut by the sawyers to the slope required by the moulds, and taken by the labourers to the side of the ship-way where the ship is to be built. The shipwright foreman of new work next sends the shipwrights in gangs to put up the frames. Every inspector of shipwrights has three gangs of men; each gang has a leading man, fifteen or more men, and two or three apprentices. The first man in each gang is considered the best man, and aft hand; the second in the gang the next best man, and fore hand; the next the second from aft, the next the second from forward, the next the third from aft, and so on. The after end of the ship takes precedence in all things: the senior inspector, for instance, takes the starboard side aft, the next the port side aft, the next the starboard side forward, &c. By this system every man knows the station on the ship where his work should be without any other appointment; but the folly of unpractical and ignorant officers to a great extent confuses the work by taking men from their proper station to work on jobs which should be done by other workmen. If the men are allowed to do the work that falls to their station it is generally done well and quickly, as those who choose, or, as it is called, shoal the men, take it in turns for the choice of the best men, till the good men are all chosen. Then the rest of the names are put into a bag or hat, and as the names are taken out so they have to stand until another shoaling the following year. All the men are supposed to be efficient shipwrights—that is, men who know how to do any kind of work required about the hull of the ship, even to the fixing or fastening, plumbers', smiths', and blockmakers' work, and work on framing, planking, rudder-making, stern and head finishing, store-room fitting, or anything whatever. The shipwrights do nearly all the work, even simple labouring, for the labourers are seldom employed on board a ship after the frames are up. Shipwrights' apprentices, and sometimes the shipwrights, fetch stores, and are often employed to clear their own chips out of the ship. They bore their own holes, drive their own fastening, as well as chip, rivet, and cut iron work for knees, bolts, &c.; sometimes they do the caulking and painting, always do the metal sheathing whatever it may be, with iron, copper, brass, lead, or sheet tin, and they always make their own staging, as well as the staging of all others working on the ship. In fact, a great number of the shipwrights are masters of their craft, and as workmen are practical, though on the whole not very efficient men at doing any kind of work which can be required on a ship from truck to keel."

<small>The power vested in the Controller.</small> The power vested in the Controller for the indulgence of whims in the con-

struction of ships has been spoken of before, but the terms in which the functions of that official are named in the remarks of the dockyard officer are sufficiently suggestive, and the evil sufficiently serious to justify recurrence to the subject. In the Controller the whole construction of the navy is virtually vested, and for the office no professional qualification is required. Naval Lords of the Admiralty with the wildest notions of shipbuilding have only to talk him over, and he is the servant of those Naval Lords; he is himself a naval officer, and there are few naval officers whose heads are not full of fleas; those about him may influence him; and those not about him may approach him by Court or other channels. The office is not only an anomaly, but it is one that is a scandal. It is a sinecure, for a man's coat and boots may fill it; it is a position in which a man may fortify himself by concessions and sanctions the most venal; it is an opportunity that a bad man may turn to great personal account. A Controller of the Navy, if there should be such a man, ought to be one of great attainment, less a sailor than a man of business, and more accustomed to direct skilled and unskilled labour than to theorise about ships. Shipbuilding is the least important, and always will be the least important, of his duties, while the wise expenditure of the large sums voted annually by Parliament should be his constant and untiring care. Of influence he should know and recognise none but that of trusty councillors, who should be about him, and by whose judgment, particularly as regards ships, he should alone be guided. Those gentlemen in and out of Parliament who have asked for the establishment of schools of naval architecture must first remove the Controller of the Navy, or reconstruct the office, if ever any good is to come of such schools;

for, as things are at present, the Controller hears, sees, and knows only what he likes or his superiors fancy.

<small>The peculiar duty of the Constructor.</small> If the office of the Controller is a sinecure, that of the constructor is the contrary. The Controller orders; the constructor executes, on paper. It is the case of the country tailor and the country customer. A suit of clothes of a particular make is wanted, and here is the corduroy. The constructor looks first at the corduroy, then at the Controller, all the while scratching his head thoughtfully. He sees his way to a coat and trousers, or to trousers and waistcoat, but beyond that he is at a loss. The Controller asks him to bethink himself, as he is sure the corduroy is ample for all his wants. No, it is impossible; and the constructor succeeds in time in showing that it is. The Controller then withdraws to think the matter over. He will now confer with the Naval Lords whose instrument he is; or if the crotchet is his own, he will put it in another and more likely shape; and so on. With the constructor he in time seeks another interview. A particular form and class is wanted, and the constructor is at liberty to use all the cloth he pleases. Is this a system to advance the science or the art of shipbuilding? Is this a practice to promote the interests of the British navy or to maintain the influence of British power in Europe? The constructor of the navy should be a man to direct and counsel. The country should look to him for the rescue of shipbuilding from the state of scientific degradation into which corruption and incompetency have suffered it to fall.

<small>The little that the Superintendent has to do.</small> The Admiral, Commodore, or Captain-Superintendent receives the order for a

ship, together with the drafts and specifications, from the Admiralty, and delivers them to the master shipwright. The duty reads like a joke. Why not send direct to the master shipwright? To do so would be to supersede a great functionary. Were the Admiralty to communicate direct with their heads of departments, a wanton and irreparable injury would be inflicted on a meritorious and important class; in other words, a very considerable back-stairs prop would be withdrawn from our naval system. The command of a dockyard is an employment to which all distinguished naval officers of rank have a well-established and proper claim. Change, therefore, would be injustice. And are not the Superintendents most deserving? Do they not supervise everything and see to everything? Step into the dockyard of a morning, or of an afternoon, and is the Superintendent not about, watching the public interest? When he approaches, are not the idle diligent, the careless attentive, and the well-behaved sure to be encouraged? The answer must be that the Superintendents are a nuisance. In the dockyards the most willing and conscientious can do no good, because they know nothing of dockyard matters, and the meddlesome, conceited, and whimsical are always doing mischief. If the Superintendent represents the Admiralty and is a check on every one, he is himself unchecked, and all experience proves that he stands in more need of watching and restraint than all the thousands that may be under him.

<small>The red-tape routine before anything is done.</small> The amount of red-tape routine before anything is done seems almost incredible, but nevertheless is true. Form is carried to such a length that it stops short only at workmen not being required to get the written permission of some official or other before

attending to the calls of nature. Men may walk, stand, sit, chew, and the like, without the formal sanction of those in authority over them, but they can venture on nothing more. Everything must be done according to strict rule or it must remain undone. Take a case. A party of shipwrights can do nothing until a handful of tenpenny nails are driven. In a private shipyard, one of the party would at once put on his jacket and be off to the office in a twinkling for them. Not so in the dockyards. The foreman has gone to the other side of the dockyard about some timber, and although the whole party of shipwrights were to go in search of him the chances are that he would not be found. Well, until the foreman turns up no nails can be got; he being the proper person to apply for them to his superior officer. But suppose the foreman is at hand and as anxious about the tenpenny nails as the men. Then it may happen that the proper officer to apply to the storekeeper is out of the way: it may be for an hour or for the afternoon. This is the too frequent operation of the routine of the dockyards. Those who framed the red-tape code were not men of business, nor even can it be said of them that they were men of the world, because a very slight acquaintance with human nature would have sufficed to teach them that working men are, as a rule, more upright and trustworthy than those above them. All experience proves that no obstacle whatever should interpose between workmen and the tools and material of their labour. No doubt instances of petty pilfering will occur sometimes, where there is no restraint, and no other supervision than that exercised by the time-keeper at the gate; but the dockyard system has not yet stopped pilfering, and apart altogether from the delay and inconvenience and consequent loss routine occasions, it has the bad effect of making honest workmen feel that they are thought untrustworthy.

The unwieldy Gangs. The unwieldy dockyard gangs are the original of the unwieldy convict gangs, which have been so much and justly censured. A small job which in a private shipyard would not keep more than half a dozen men going is in the dockyards usually graced with a gang eighteen or twenty strong, because it does not answer to break the gangs. The half-dozen men who might do the work would be without that supervision which the strict letter of the instructions of the Admiralty peremptorily demands, and no doubt would take advantage of the chance presented to them of either doing little work if on day pay, or of charging twice if on task and job. So the whole gang go to perform the work of six men; and to prevent the men from robbing in a small way, the public are robbed wholesale by what in effect is the compulsory idleness of twelve or fourteen men for a day, week, or month, as it happens. Visit St. Mary's Island, Chatham or Portsmouth Dockyard, where the convicts are at work, and a score of able-bodied men are drawing along an empty hand-cart; doing the work of one of their number only, because where one goes the whole gang must follow. Precisely so is it in the dockyards among the dockyard workmen, although there are no hand-carts in use to admit of the unwieldy gang system appearing in its true colours.

Shoaling; the farce.* Shoaling the dockyard shipwrights

---

* It may be well to state that skulking fellows in the dockyards get surgeons' notes for easy work, which enable them to get into easy places. Doctors render the same kindness among the convicts.

is a farce. After all the best working places have been filled recourse is had to the ballot. The names of the unfortunates are written on slips of paper, and the work on which they are to be employed throughout the year is fixed by lot. Qualification is ignored. The best workmen may get the roughest work, and the converse, but what matters it? Whose loss is it? The loss is that of a long-suffering public, who, it is said, would complain if the system did not suit them. Among the workmen it is a common saying that the loss, if any, comes off a broad board: the meaning being that it hurts no one in particular very seriously. And yet Parliament is frequently assured from the front benches, on the warm side of Mr. Speaker, that the workmanship of the dockyards surpasses that of the shipyards. It might as well be told that the savoury atmosphere of the Thames in July is more grateful than the breezes on the coast; or that the boot-making, the smithing, and the carpentering of London thieves on St. Mary's Island, Chatham, is better than the honest workmanship of the honest tradesman. Then the ballot sanctioned by the Admiralty is indeed a wrinkle for the society in Guildhall Chambers and the advanced Liberals. Who ever thought that under the antique rule of the Commissioners for executing the office of Lord High Admiral there lay concealed this choice morsel of genuine Yankeeism? But, after all, are not the Commissioners themselves merely shoaled into office? no fitness for their position being recognised, although the slips of paper and the hat or bag are not visible, unless in the chamber of the successful party leader.

The usage of precedence. What shall be said of precedence and grey hairs taking the post of honour aft among the dockyard shipwrights? The usage is a monstrous one. What would become of a private firm if the old fellows, instead of taking hold with the other workmen, were privileged to stand next the counting-house or to single out easy tasks for themselves? The firm might as well frame a pension-list. It would be supporting what are known as deadheads. It would be running the hard race of competition with so many mill-stones round its neck. But there are no such private firms to be found, nor are there any such old fellows outside the dockyards claiming the current rate of wages. Go into any private workshop where old men are earning as much as young men, and you will find that they *are* earning it. They are visibly at war with age, resolutely and successfully contending with it. Instead of being a drag and bad example, they are positively a good example and benefit. Young fellows never see withered arms exerted vigorously, nor withered forms bending under heavy loads, without being fired with emulation and exerting themselves in a worthy manner; and if this activity of the old exercises an influence so potent, what is to be looked for in the dockyards, where every old fellow is allowed to skulk? Either old men should be turned out of the dockyards, or they should work as young men. In business there is said to be no friendship, and in trade there should be no favour, for labour is practically as much a commodity as the product on the creation of which it is employed. Precedence among dockyard shipwrights is a quiet way of robbing the Exchequer.

The unpractical Officers. The unpractical officers who remove men engaged in shoal work do something more than break the gang rules. They often leave the men's work undone. Strict rule requiring the men of a gang to stick to their individual tasks, the absent men's tasks remain untouched while the adjoining work proceeds. Take an illustration from the building of a house. There are twenty men on the front, with three feet each to raise and finish, and the four centre men are withdrawn to re-floor the Superintendent's garret or pack his furniture for the moors or Brighton. The consequence is that the ends of the front of the house are built up while the front door and the immediately adjoining parts remain untouched until the absent men return. So with the planking or other work on a ship. A third or fourth of the gang are taken off and the work of those men lags behind, to, it may be, the serious detriment of the ship and to the certain injury of the gang if they are on task and job. Should the gang be on day wages, the ship, of course, is the only sufferer, because the men can take things easy until their mates return. With the unpractical officers no one, of course, dare remonstrate, and the men suffer silently the wrong too frequently inflicted on them.

The Jack-of-all-trades character of the Shipwrights. The Jack-of-all-trades character of the shipwrights may to some extent have fitted them for the change from wood to iron. Before accustomed to various kinds of metal work, the tools are not altogether new to them; but between wood and iron shipbuilding there is practically as much difference as between the railway and the stage-coach. Men could handle timber at their leisure, and drive trenails home as lazily as they pleased; but to rivet the bottom of an iron ship, and fasten the plates on an iron ship's sides, men must strike with the iron hot or there is no use striking. They must wake up to iron ship work, and that is what stiff-jointed old wooden shipwrights never can and never will do. And it is the case that men who have all their lives changed employments, as the dockyard shipwrights have done, cannot contentedly apply themselves to one pursuit. They cannot fall willingly or well into the ranks as furnace men, frame binders, angle-iron smiths, platers, riveters, drillers, clippers, caulkers, &c., and nothing else. The Admiralty may say they do; if so, the proof is wanting at Chatham. There the old wooden shipwrights are worse than raw clodhoppers in uniform. The blacks are obnoxious to them, they are not quick enough, and the endurance essential to good iron shipbuilding is too plainly wanting.

<small>The admission of inefficiency.</small> Last of all, the admission of inefficiency reflects credit on one wearing the dockyard livery. How could it be otherwise? What is there in the entire dockyard system to place dockyard shipbuilding on a level with that of any foreign country, still less on a level with the private shipbuilding of this country? The Controller ties down all to his iron will, and the constructor, who should be his master, is his lacquey. The Superintendent too often is a fool; the master shipwright's sole occupation is to sign his name, and any one dressed in authority may interfere with work and workmen at their pleasure. To look for efficient workmen under such conditions would be to look for calms round Cape Horn. The system renders good workmanship impossible. It is also calculated to ruin the best-inclined and competent among the shipwrights, who are enticed into the service in the hope of being superannuated in their declining years, whether they

have been provident or improvident with their weekly wages.

Shipyard Shipbuilding. The Admiralty, so far worsted in argument, may, however, take refuge in the assumed defects of shipyard shipbuilding. They may take to throwing dirt at the private shipbuilders, in the hope of making out their case for a continuance of dockyard shipbuilding in the form in which it is carried on at present. To defeat this, let us glance for a moment at the salient points of shipyard shipbuilding.

Designing Merchant Ships. With regard to the designing of merchant ships, it may be said that in nine cases out of ten British shipowners and their nautical advisers, by whim or fancy, determine both the form of ships and the disposition of the masts and sails; that they systematically reject the counsel of those trained to know and apply the truths of science in such matters, and that the great merchant shipowners usually have no two ships alike either in hull or spars. Let us admit the general truth of these statements, for there is more or less of truth in them, and what conclusion do they logically suggest? Certainly not that, because these things are so, dockyard shipbuilding is all that can be wished? And as certainly not that the designing of merchant ships is for these reasons in the least objectionable. Why should not the owners of merchant ships experiment in forms and in the disposition of masts and sails, even although by so doing they should be open to the charge of not being held in leading strings by those who have made shipbuilding their profession? If they reject professional advice, it must at least be said on their behalf that they have much more at stake in the course adopted than the professionals. They have ships as an investment for their capital; naval architects do no more than design ships to illustrate shipbuilding science. So this objection falls to the ground. Irregular as the conduct of the ship-owners may appear, it nevertheless is proper, although it is not unworthy of their consideration whether, as a rule, it would not be better to confide in a greater measure the design and construction of merchant ships, spars and sails included, to those in whom the trust may properly be confided.

Contracts with Workmen. It may be charged against the shipbuilders that there is no uniformity in the hiring of workmen, for whereas in some yards the men work by a price-book, or stated price for stated work, in others all are paid by the day; and in addition, that in several yards " an agreement is drawn up by the master or employer, and signed by the man or men to be employed and the employer—this agreement being retained by the employer—who seldom gives a copy to the man or men engaged, so that the men are bound or not bound just as it suits the purpose of the employer." While all this may be true, the charge, however, does not represent the whole truth. The fact is that in all the great shipyards of the country there is absolute uniformity—the uniformity of contract pure and simple—the workmen earning just as much as they can, and that as the minor shipyards assume larger proportions, no choice is left to them but that of falling into the custom of the great shipyards. Then as regards the breach of contract and the overreaching of the workmen spoken of, the truth is that such a practice is positively unknown, and positively repudiated by every firm with the least pretension to respectability. Nine in every ten of the shipyards never heard of such a thing; but let it be assumed that the tenth is a guilty party. Well, what can this prove? Why, nothing more than that there are tricks and dishonest people in all trades. By no rational stretch of imagination can it justify dockyard officials, or Lords of the Admiralty, deprecating the terms of employing shipyard labour, and praising the terms of employing dockyard labour.

<small>Antagonism between Employer and Employed.</small> Or it may be charged that while perpetual harmony exists between the dockyard authorities and the dockyard workmen, frequent jars and strikes take place in the shipyards. Let us again admit the truth of this. Is it, then, to be said that because the dockyard people pull the horse well together that no fault is to be found with the dockyards? It might just as well be said that because goodwill existed between the directors and clerks of the Royal British Bank, or between Messrs. Strachan, Paul, and Bates and their clerks, these establishments were models of propriety. Or we might say that because the convicts and warders in Milbank Prison enjoy the fat of the land in peace and contentment, that convicts are not a bad set of people after all, and that leather medals might be bestowed on the warders. The jars and strikes that take place in the private shipbuilding yards are perfectly compatible with the best working system that can be devised. Workmen resist some masters because they can get employment elsewhere on what they conceive to be better terms—that is, less onerous. Workmen also strike because they conceive masters ought to yield certain things. In the antagonism between employer and employed there is consequently

nothing that is unreasonable or improper, although it has often been impossible to help deploring the loss arising from strikes, and the sufferings of wives and children. But jars and strikes are certainly reduced to a minimum under the honest, encouraging, and profitable contract system. Here is a certain amount of work; what will you do it for? If the workmen are not thoroughly incompetent, they will have nothing to complain of when their own terms are acceded to, or when they accede to the terms of their employer. Misunderstanding ought to be impossible, and the workmen must be sensible of the dignity of their position and the advantages of maintaining it. Jars and strikes in shipbuilding are, however, inseparable from the day work and day wages system, particularly when the employer is of that antique pattern which adorns the dockyards. Such a man is never pleased, but always growling. His motto is not "Live and let live," but waste soul and body unprofitably in his ungracious service. This type of shipbuilder may still flourish in unimportant country towns, but he is seldom met with in the great seats of British shipyard shipbuilding.

Chapter V.

NAVAL POWER.

The generally assumed elements. Seamen. The generally assumed elements of naval power are dockyards, ships, and seamen; dockyards first, next ships, and last seamen. It was only the other day that the first attempt to organise reserves of seamen was really tried, and inasmuch as only 18,000 of the 40,000 which it was proposed to enroll have yet been embodied, the experiment is a failure. It is a failure, and a very melancholy one, when viewed apart from the crowd represented by 18,000 fine young sailors. Eighteen thousand from forty thousand is in commercial slang something like seven-and-six in the pound, and no one will say that at any time the figure is much to boast of. Why the enrollment has failed is because the Admiralty, always generous to profusion to unworthy people, and for unworthy purposes, is shabby to the sailor. £600 per annum to a master shipwright, with house, coal, and candle, is regarded as so much below the mark that when the master shipwright thinks proper to retire he will have £300, £400, or £600 per annum, and an equivalent for house, coal, and candle, all his days. And all that a master shipwright does is to sign his name. The sailor who is to defend our firesides is offered the inducement of £6 per annum. Nothing but patriotism could have led a single man of spirit to accept the terms. £10 should have been the minimum gratuity to able seamen, and £15 to mates and masters irrespective of the pay and allowances during the annual training. This is the opinion of a great number of the men, and the all but universal opinion of the unenrolled. As it is desirable that the men should speak out and the force be raised to 40,000 as a minimum, the best thing that they can do is to memorialise the Queen, and if their demand is not acceded to, to lay aside their caps and blue shirts. Englishmen do not want their reserve sailors to be underpaid, and will support them in their just demands. To take a place in the marine reserve should be as much worth the while of the men as it is of the country, and there is surely sufficient administrative genius among us to provide 40,000 seamen with ten-pound notes each out of the £250,000 weekly, the £1,000,000 monthly, and the £12,000,000 annually voted and spent for the navy. The easy way in which the Admiralty take the indifference of the mercantile marine is in perfect keeping with the narrow views entertained by My Lords on the subject of naval power. The Admiralty are, and always have been, pestered with a superabundance of seamen, unless during the Russian war, when the supply was so much short of the demand that Sir Charles Napier was expected to take Cronstadt with a fleet contemptible in numbers and equipment, and half, if not three-fourths, manned by young offenders from reformatories, street pickpockets, and assistants and servants out at elbows and out of place. Just now, Portsmouth, Plymouth, &c., are positively running over with unemployed navy seamen, and why should the Duke of Somerset and his colleagues care a straw whether the seamen of the mercantile marine are pleased or not pleased? Have not their lordships quite enough to do bolstering up the system of good Henry VIII.'s time, with its sinecures and great dockyard almshouses?

<small>The superior claims of Ships.</small> Ships have always had superior claims to seamen with the Admiralty. With ships there was the genial flow of patronage, while with seamen there was nothing of the kind. What scion of a noble house, what partisan, what constituent could be propitiated by the nomination of a second-class boy, the appointment of a coalheaver or stoker, or the entry of an ordinary or able seaman? Seamen, therefore, are and always have been beneath the notice of My Lords. Nothing could be made of them, so seamen in the British Navy have really and truly from first to last been deemed a nuisance and kept down. Not so with ships. They have always yielded an official harvest. They have always been a bribe to the dockyard towns, a means of obliging no end of people, and of weakening the ranks of the Opposition when the Administration felt its hold on office feeble or insecure. Constructing ships of war has been as much the hobby of aged Captains and Admirals, as amalgamations, extensions, and oppositions are of railway directors now-a-days, or breeding stock of modern agriculturists. A whipper-in could always, and will always, buy over an Opposition Admiral by offering him, directly or indirectly, a finger in the shipbuilding pie. Then in connection with ships, either in construction or commission, there were always desirable appointments turning up, which all might ask, and none blush to take. And last of all, ordering the construction, the fitting or commission of ships at particular dockyard towns, has been an act as definite and intelligible as the ordering of a Christmas dinner in a

workhouse, or an extra distribution of quartern loaves on Sunday in famished Lancashire. The building of the *Achilles* is a bribe to Chatham, and so will be the building of the *Lord Warden* and the *Bellerophon*.

<small>The surpassing claims of Dockyards.</small> But if ships have had superior claims to seamen, dockyards have had, and still possess, superior claims to ships. Seamen are the menials, ships the emblems, and dockyards the repositories of the greatness of the Admiralty. My Lords, careless about seamen and undecided about ships, would guard the dockyards with a flaming sword. Their minds are perfectly made up about dockyards, knowing as they do where and on what side their bread is buttered. Sooner may their right hands forget their cunning, and their tongues cleave to the roof of their mouths, than that the dockyards should fall a prey to the French. The construction of Keyham was an enormous bribe to Devonport, the construction of the superfluous and costly tunnel from Devonport to Keyham was a second enormous bribe to Devonport, and the extension of Keyham and the filling of the factory with machinery is a third enormous bribe to Devonport at this very moment in serious contemplation. The reclamation of St. Mary's Island, Chatham, from the Medway, for the purpose of erecting iron-foundries, rolling-mills for armour-plates, and boiler and engine shops, is an enormous present bribe to Chatham, and the extensive defensive works recommended by the Defence Commissioners on Chatham Heights is an enormous bribe for a later period. The defences at Spithead and Portsmouth can be viewed in no other light as regards Portsmouth. In fine, the dockyards are the first and chief care of the Admiralty, and you will be told that we have only to fill and fortify

them to bid defiance to the world. While ships and sailors pass away, the dockyards remain and are eternal.

<small>The obvious inadequacy in Dockyards.</small> The inadequacy of the Admiralty notion of naval power admits of being established in very few words. As regards the dockyards, it never has been pretended that Sheerness, Chatham, and Woolwich and Deptford would do more than repair the casualties of a hard-fought action in the North Sea; or that Portsmouth, Devonport, and Keyham would do more than render a like service in the Channel. Pembroke, as is well known, is a mere building-yard. The dockyards, therefore, in their most imperfect way, would provide for casualties round the coast, from the Nore to the Land's End. Is that a sufficient provision for a great country? Is that an adequate insurance, as it is called, that our shores should remain inviolate? No man in his senses will assert that it is. If dockyards are the good things they are said to be, one at least is required on every headland. Suppose the enemy's fleet overtaken in the Frith of Forth, in the Pentland Firth, or off Holyhead, and an action fought; of what use would Portsmouth and Co., and Sheerness and Co., be to the disabled ships in such a case? Why, they might as well be at the Antipodes if modern gunnery is only half as destructive as we are led to suppose it will be. Or take another case: if the enemy were chased into the Atlantic, and an action fought with steam supplied by stores and woodwork, what would be the use of Portsmouth and Co., and Sheerness and Co., if the disabled ships were compelled to make the land on the north coast of Ireland, or at the Orkneys? Surely there might as well be no Portsmouth and Co., and Sheerness and Co., as far as such

ships were concerned. These ships would require to put into the first roadstead, and get into fighting trim without fuss or loss of time, if the officer commanding were to discharge his duty in a manner that his countrymen would approve. These are considerations that appeal to the common sense of all: if dockyards are necessary for the overhaul and refit of a disabled fleet, the miserable handful of dockyards that we possess is inadequate; and if dockyards would in most cases be utterly useless because unavailable, so they may be dispensed with in all cases. Logically the dockyards should be numberless, or there should be none. But softly it will be answered, why not to a moderate extent avail ourselves of dockyards? There certainly can be no objection to making the most of the existing dockyards for the overhaul and repair of disabled ships, but in the name of reason let it not be pretended that the existing dockyards will be any great reliance during war. A good shipyard in the Pentland Firth might on a pinch render as acceptable and good service as Portsmouth and Co., or Sheerness and Co.

Comparison with France in Dockyards. To a numerous class accustomed to accept opinion on trust in such matters there will, of course, remain the very great consolation that, be the state of things what it may, England, if any, is not far behind France in the number and greatness of its dockyards. Those to whom this reflection yields comfort are rejoicing in a fool's paradise. A comparison of the number and resources of the dockyards of both countries suggests nothing, teaches nothing, and ought to frighten no one on either side of the Channel. Commercial men, looking forward to a great financial crash, such as those which swept over

America and Europe in 1837 and 1857, might as well attempt to reckon, in advance, the stability of Paris by the returns of the Bank of France, or the stability of London by the returns of the Bank of England. The Bank of France is not Paris, neither is the Bank of England London; and so the dockyards of France or England, be their number, extent, or resources what they may, are not a proper standard by which the building, repairing, and fitting capacity of France and England, as regards ships of war, should be measured. We might just as well take the coats on people's backs as the criterion of character, position, or of means. The building, repairing, and fitting resources of France are immeasurably in excess of its mere dockyard capabilities, and the same is true of England. Is it not, then, an insult to intelligence for England to taunt France, or for France to taunt England, on the subject of dockyards? Is it not an unpardonable weakness for any French or English Government to lend itself to rivalry in such a matter? Thus it is a small and worthless crumb of consolation for any one to reflect that after all England, if any, is not far behind France in the number and greatness of its dockyards.

<small>The obvious muddle about Ships.</small> As regards ships, the Admiralty are, and always have been, in a muddle. The prevalent opinion among Lords of the Admiralty and Admiralty officials is, and always has been, that ships are ships. Up to this point they all see their way clearly. This is the *Constance*, a noble frigate. This is the *Enterprise*, a gallant little ironclad. This is the *Minotaur*, a magnificent ironclad. Such is the fustian of official circles. Give us ships and we will fight the devil. Build, reconstruct, establish schools of naval

architecture, and afterwards damn Frenchmen and the Yankees. What on earth is to be made of this;—these high-sounding expressions and idle boasts? Suppose Americans and Frenchmen to indulge in the same strain, and the vanity and absurdity of the proceeding will at once appear. Neither boasting nor personal congratulation ever won a battle. Boast as we may, and praise our ships as we like, the race is not always to the swift, and unless the construction of ships of war proceeds on some sure principle, it is probable—nay, certain—that when the day of battle comes we shall, as before, miss the mark. Before the outbreak of the Russian war we had been building and accumulating ships in various ways for half a century, and it turned out that we had plenty of ships that were not needed, and not one of the class that was urgently required. For years members of the Board of Admiralty had been puffing the navy at Lord Mayors' dinners and other exhibitions: Englishmen of all classes were labouring under the delusion that we held the sceptre of the sea; and yet, let the honest truth be told, the navy was only getting into a condition to do its work when the Russian war was over. The Admiralty were muddled, and the service was muddled. Both are muddled still. Were we to become again involved in war with Russia, the ships are not on the Navy List, nor in the public or private yards, nor even in the heads of officials, that the occasion would demand, and our bunting would be again disgraced before Cronstadt and Sebastopol. So it would be in a war with the United States. For the requirements of such a war, not a single angle iron has yet been rolled, nor a single log of teak sawed. The Duke of Somerset can only talk to you about the activity of the Chatham workmen on the *Achilles*, and of some mare's nest or other that Admiral Robinson has just discovered.

<small>Comparison with France in Ships.</small> But it will be said, in the complacent way that is so persuasive to weak minds, France is the only nation that we have to fear at sea, and we are on the point of enormously outstripping that country in ironclads. Indeed, it will be added, our *Warrior* is as good as any two French ships; while our *Minotaur*, our *Northumberland*, and our *Agincourt* will be more than a match for the ironclads of France and Italy, should both ever pull the horse together. This, a moment's reflection shows, is the logic of the dockyards. An incomplete and therefore erroneous value is set on the ships of France, and our own ships are overrated. From such premises the conclusion follows that the French ships would be swept from the Channel and the sea. All with the least pretension to scholarship will at once concede that this is the reasoning of the kitchen, the nursery, or the madhouse. Yet it is substantially the stuff that Lord Clarence Paget is in the habit of serving up to the House of Commons with the Navy Estimates. He runs over the list of the French ships and the list of our ships, and timorous gentlemen mentally thank God that the country is better off than they imagined. Now, is it conceivable that France and England will eve rengage in naval war, or rather, will ever fight naval actions in the same way as people play cricket, or as Oxford and Cambridge row matches at Mortlake? It is incredible that they ever will. Half a dozen well-coaled French ships or half a dozen fast French ships may batter to smithereens a dozen coal-exhausted or slow English ships; and the converse. Of what value, then, are numerical computations? Again, half a dozen French

ironclads may dash up the Thames, the Mersey, the Clyde, or Tyne, while a dozen English ironclads are busy looking out for them in the middle of the North or South Atlantic. Or, last of all—for it is needless to multiply illustrations to the same effect--half a dozen French ironclads might temporarily draw off the whole of ours to protect Gibraltar, Malta, &c., or in the expectation of fighting a great and decisive action on some historic battle-ground, while another half-dozen, let us say, covered the landing of a force in Ireland. Thus comparisons with France in ships may alarm nursery-maids, but they should not alarm others. Strong-minded men should be above the weakness. Not the number of ships that may be possessed or that may be brought into action is to be feared, but the skilful manner in which they may be fought and the readiness with which they may be sent to sea again after being disabled.

The obvious error about Seamen. The Admiralty error as to seamen is sufficiently obvious almost to be passed over without further notice. But there is one remarkable fallacy, to the exposure of which too much prominence cannot possibly be given. We have a mania for ships, and still are so indifferent about seamen* as to exclude them from the quarter-deck, to flog them, and to pay and provide for them as if they were hodmen or

---

* Of late years the treatment of navy seamen has been impolitic and unjust. From 1700, and probably for a century before that, the officers of the navy ranked and took command in the following order:—Lieutenant, master, second master, boatswain, gunner, carpenter. Master's mates and midshipmen were then ranked as quarter-deck petty officers. About 1832 mates took precedence of second masters, and consequently stepped over the heads of the warrant officers; but the finishing blow to their rank was given in 1844, when they were placed below "subordinate officers."—*The Disabilities of Royal Naval Seamen;* by Joseph Allen, 7, Wellington-street, Strand; 1863.

dockyard scavelmen The truth is, that our seamen lead a life more truly wretched than the Carolina negroes used to do. Under a gloss of light-heartedness, most seamen are consumed with care. Their pay, ample for all the proper wants of boys and single men, is insufficient for the married. Enter the homes of such, and while wretchedness stares you in the face, the traces of vice are apparent to a moderately close observer. The half-pay of the husband, under the most skilful handling, would just suffice to keep the soul and body together of wife and little ones, and under moderately unskilful spending cannot do so. Some wives resolutely betake themselves to honest callings: char, wash, iron, or white work; but many do not. The great majority become immoral, then dissolute. Their children swell the ranks of crime; the honest, manly sailor, then undone, goes from bad to worse, and in time is dismissed or leaves the service. Now, what are seamen to the State that they should be the victims of this neglect? They are to ships of war what skilled workmen are in manufacturing industry. Without them ships of war are useless, and possessing them, without possessing ships of war, enterprises of pith and moment might be engaged in, in the presence of an enemy. Foremost among the laurels of the British navy will always stand the cutting out of ships with boats, the destruction of booms, the towing of fire craft, and the storming of forts and works. But for the daring and endurance of our seamen, not on one but on all occasions, and under all circumstances, that prestige which we cherish and the world trembles at would not be ours. Nevertheless it is the fashion to cherish ships and neglect seamen. At this moment, and during many years, we have possessed infinitely more ships of war than could by any possibility be manned by even half—nay, quarter-trained seamen. The State, therefore, as regards ships, acts the part of the madman who would spend his fortune building houses which none would occupy. Reserve forsooth! Surely the first condition of a reserve of ships of war worth possessing is our ability to man the ships on the instant, not with the scum of London, its pickpockets and juvenile offenders, but with men who have sea-legs, who are practised in the use of small arms, accustomed to authority and submission, and, above all things, taught to use shot and powder with alacrity and effect. Just now, and at all times, did it suit the purposes of France, the great and well-organised reserves of that country might literally swallow ours.

<small>Comparison with France in Seamen.</small> Every word applicable to the case of ships is necessarily applicable to the case of seamen. As with ships, there can be no comparison of seamen of any real value.

<small>The real elements of Naval Power. Readiness.</small> The real elements of naval power as contradistinguished from the supposed elements are few and intelligible. The first is readiness; the second, resource; and there is a third, which may be separated from the second—endurance. A pail of water at hand when the wainscoting of an old house takes fire, might prevent a conflagration. A man, physically unable to cope with any one, making his way home at night through a suspicious neighbourhood, and stopped at a street corner, would settle a fellow, who otherwise might strangle him, by a single well-directed blow with his walking-stick, if the blow were instantly delivered. Often does a mere stripling knock down and punish a bully twice or thrice his weight by the exertion of his

relative feebleness in a scientific manner, taking his adversary unawares. Readiness, therefore, may prevent no end of mischief, and with or without scientific attainment, may enable the weak to overcome the strong. The rule is of universal application; true of all times and all occasions, both on land and sea. The lamentable American war has furnished many instances on both sides. Stuart and Jackson, with their little bands, inflicting all but irreparable injury on their powerful and more unwieldy adversary, are the most notable. In the glorious Italian war, Garibaldi supplied numerous examples of the same kind. *James's Naval History*, to leave the present, abounds likewise in deeds of successful daring; great fleets scattered by small fleets, the former taken at some disadvantage; roadsteads surprised, and so on. To readiness in modern war belongs the foremost place. It is worth battalions in the field, and squadrons on the ocean. A ready nation, up to a certain point, has success before it; and an unready nation, chastisement and disgrace. Let us represent to ourselves ready France and unready England engaging in a naval war, both being opposed. France, resolved to profit by its advantage, might either bide its time until preparation could be pushed no further, if the British Government appeared supine, or it might utter the ill-omened word at once if the British Government vigorously hastened the equipment of the fleet. The consequences would not fail to be calamitous to us, should the French Admiral perform his duty. No doubt our unreadiness would in time be rectified; but the point is that the issue of a dash might possibly be overwhelming to the stronger. On the other hand, were both nations ready, and surprise on either side impossible—as far as impossibility can be thought of—

a new and probably a sufficient guarantee would exist for the maintenance of peace.

<small>Effective readiness.</small>   Effective readiness now-a-days is in that measure which accomplishes an immediate purpose, and it is to be taken in connection with that new condition of modern war—the opinion of the public. When France suddenly stopped short at Solferino, it was because its preparations and calculations did not go the length of war with Germany; its readiness up to the very point at which it stopped was effective, but no further. Some months ago, it will be remembered, during one of the many crises in our relations with the American Government, one of the members of a deputation of trades unionists waiting on Lord Palmerston, remarked that were this country to go to war with the United States, civil war would break out among ourselves. This was evidently an unwarrantable assumption on the speaker's part, but it shows the influence of the educational agencies now at work, more particularly the penny newspapers; working men now venturing to take sides with those whom they imagine to be right. A similar display of working-class sentiment was very generally remarked during the *Trent* affair. Thus, to speak of ourselves again, and the occasion of a war with France, the course of events would be watched and influenced by public opinion in a manner that would startle some people. During the long war, and for a considerable period after, freedom of speech and freedom of writing were imperfectly enjoyed, and for that reason the ruling classes had everything pretty much their own way. That time has gone by. An English Government entering into a war with France which the public opinion of the country only sanctioned by a bare class majority,

would be in danger of having peace thrust down their throats if the Solferino dash were repeated in the Channel. Some may smile at the apparent simplicity of this conjecture, but they are perhaps not aware that among the more advanced Radicals there is an impression gaining ground that no moderate Parliamentary changes, such as Mr. Bright used to advocate, are likely to be obtained until the country suffers some great humiliation by war as in Canada, by rebellion as in India, or by invasion from across the Channel. They refer you to history and affirm that the spirit of Conservatism now abroad is identical with the intolerance that reigned after Waterloo, and that the Reform Bill, the repeal of the Corn-laws, and other minor changes were the fruits of Court humility, if not of terror. Hence the reasonable possibility of even a Solferino truce on English soil. This at least is certain, that were we to become involved in war with France, it would not be such a war as the last, bloody, purposeless, and protracted, but one to which the enlightened public opinion of the time would assign limits. For France, therefore, to attack this country, its preparations need not assume out-of-the-way proportions as when Napoleon I. meditated a descent; if it were only pretty clear that the policy or course of the British Government would be unpopular, and one overwhelming French dash could be made at London or some other part. Take another case. The difficulty, let us say, is between this country and Brazil; a giant is to strike a pigmy. This was a case where an act of war was committed by the British Government, and where the readiness to strike was present, but peremptorily restrained by—opinion. Had war been declared the Government must have forfeited the confidence of Parliament, and their successors would no doubt at once have engaged the good offices of a friendly Power to obtain peace. This is a most suggestive circumstance, and the influence to which it points is strongly confirmatory of what has just been stated. Vast as the warlike preparations of the world are, this is less an age of brute force than one in which there is a strongly marked tendency to redress the excesses of intemperance and ambition by the soberness of expediency and common sense. It is an age, besides, in which a wisely governed country will be ready on the instant to let slip from the leashes the bloodhounds of war, but with the mental reservation that really inconvenient sacrifices shall not be made.

<small>Organised readiness.</small> Organisation and combination are essentials of readiness. A ready fleet must be perfect in all its parts, and adapted to the varied work which active war requires. In addition to the ordinary classes of fighting ships, adequate provision should exist for conveying orders by despatch-vessels of the highest speed, and for supplying coal and provisions by tender and transport. No inconsiderable share of the success of the Confederate cruisers has been owing to the successful timing of arrivals of coal and ordnance stores in certain latitudes on the high seas, but the credit of success is mainly due to our own mercantile marine. With ships adapted to all the drudgery of attending on a fleet, this country, of course, has an inexhaustible supply in the mercantile marine, and France has so, too, but necessarily in a less degree. But in this country, where nothing is said to be organised, the Admiralty enter the market for transports, bidding against the Ordnance-office, the War-office, the India-office, and the Emigration-office. Perhaps in a moment of pressure this

extraordinary competition of Government departments would not occasion great loss in cash or much waste in time, but manifestly as long as the Admiralty are not possessed of a fleet of transports of their own, some Admiralty department should exist, charged with the merely formal duty of keeping the run of suitable merchant vessels for despatch and other purposes, and vested with authority to lay hold of those wanted on terms mutually advantageous to the owners and the public. The absence of organised readiness is, however, a trifling fault by the side of that apology for combination which has recently attracted so much attention—namely, the sailing of the Channel fleet round the coast. In Admiralty annals this cruise is a feat. It is something that was never ventured on before, at least since the construction of ironclad ships. From Portsmouth to Plymouth and back, from Plymouth to Cork and back, or from Portsmouth to Lisbon and back, are the routine rounds of the home ships. No greater peace schooling is allotted to officers and men in the Channel and in the Mediterranean and elsewhere; and it is well known that commanding officers take their duties as quietly as My Lords in Whitehall. The consequence is that proficiency in the higher duties of man-of-war seamanship is impossible in the British Navy. There is no rendezvous in mid-Atlantic which all ships must reach on a given day; no parting of the two divisions and making sail in chase in good earnest; no gun exercise when it is blowing great guns; no intricate manœuvres to illustrate practically the exploded tactics of sailing ships; and none to familiarise officers of all ranks with those inseparable from steam. This is matter for deep regret. This is a criminal neglect of proper and peremptory duty. This is inviting attack by reason of the absence of organised readiness in the fleet. It is indubitable proof that, whatever the condition of the navies of other countries in the first essential of naval power, the navy of Great Britain is on a peace footing, and not prepared to surprise an enemy.* Our naval officers are really but apprentices, because the Admiralty

---

* Last session Sir John Hay directed the attention of the House to the present unsatisfactory condition of the promotions and retirements in the navy, so that the existing state of things is most unassuring. Sir John, in the circular emanating from him, said—

It is proposed,

1. That a compulsory age retirement be applied impartially to all ranks, with an earlier optional retirement.

2. That the reserved and retired lists be consolidated on an intelligible system, and that the advantages to which an officer became entitled at the time of his retirement be preserved to him.

3. That a longer minimum period of sea service be required in the ranks below that of captain to qualify an officer for promotion to a higher active list.

4. That the system of promotion by *selection* from the ranks of commander, lieutenant, and sub-lieutenant be so far modified that *one-third* of the promotions in these grades should be given to those officers having the greatest amount of active service at sea, in their respective ranks.

5. That there be a sufficient scale of full, half, and retired pay, to enable officers to serve their country without incurring debt, and to retire with a fair remuneration for their services.

6. That there be such a reduction of the numbers on the active lists as may ensure the efficiency of officers by frequent employment.

7. That the entry of cadets be regulated according to the vacancies which may have occurred during the preceding year on the active list of lieutenants, whether occasioned by promotion, retirement, or death, and to supply the vacancies which may have taken place amongst the junior officers below that rank.

8. That a Naval College be established, with training ships attached, and that no cadets be sent to sea under sixteen years of age.

9. That to reduce the expenses, and to increase the comforts of officers and men, suitable barrack accommodation should be provided, the most important step towards the establishment of a standing navy.

10. That the position of warrant and petty officers be improved, and many of the duties heretofore performed by sub-lieutenants and midshipmen be intrusted to those officers.

By this arrangement two great advantages would be obtained.

First,—There would not be any occasion to admit into the service more young officers than those for whom promotion could be found.

Secondly,—By improving the position and increasing the number of warrant and petty officers, the prizes within reach of the seamen would be increased, and a greater encouragement thereby offered to respectable lads to enter the navy.

do not understand in what naval power consists, and officers are debarred from thoroughly mastering their profession.

<small>The second real element of Naval Power. Resource.</small> The second real element of naval power is resource. If one nation surprises another, it may turn out that a merely temporary advantage will be gained, and that the fighting must proceed. The thickest skin will then necessarily hold out longest. As in the prize-ring, the nation that comes to the scratch oftenest and most resolutely will at last win the day. The items of resource could not easily be enumerated, and the effort, if successful, would neither be entertaining nor of use. To a very few only is it necessary to give prominence. Foremost of all the naval resources of a nation is the ability to construct, equip, and fit ships for sea. The ability, the mere ability, be it observed, is the point, because stern war whets the ingenuity of man, and old stock of ships and the fittings of ships may practically cease to be of any use whatever. The other day the writer had the good fortune of travelling across the Channel, and in France, with a distinguished foreign officer, sent by his Government to report on the naval and military condition of the Northern American States— or rather, be it said, of the United States. The officer at the time was on his way home, and his American inquiries had extended as far West as the head-quarters of General Grant's army. He, while in Boston, before taking passage in the Cunard steamer to Liverpool, was asked to visit a factory in which the most formidable revolving musket he had ever seen was being manufactured for the American army. "What," he remarked to the Federal officer, "you are taking me

to a manufactory of pianofortes!" No such thing. The proprietors of the pianoforte manufactory, finding no profitable demand for their well-known instruments after the war was fairly entered on, bethought themselves of at least temporarily going with the times. They removed their pianos, took a contract for the stocks of muskets, and ended with the invention of a weapon which the distinguished foreign officer believes is still unknown in France and England. The new weapon, with which it is intended to arm the whole American forces, is an eight-shot repeater of great simplicity and effectiveness, which can be charged without inadvertence in a moment almost, indifferently in rain or sunshine, and in the dark or daylight. This is but one instance of American ingenuity to which the war has given rise. There are others as notable, but it is needless to proceed with the enumeration. Suffice it to state that up to the point to which the American war has yet been carried it is manifest that the weapons with which war is begun, as well as the material on hand at the outbreak, are all doomed to be cast aside. We are not now living in the infancy of science, but at a time when it has attained a large development. The genius which is continually storing our workshops and factories with improved, because increasingly effective, mechanism, has a comparatively new and unimproved field before it, when weapons of destruction come seriously to engage attention. Sir William Armstrong, Mr. Whitworth, Mr. Blakely, and others, in the peace that our happy country now enjoys, have neither half the stimulus nor a fraction of the competitors that they assuredly would have were we to become involved in a life and death struggle such as that which now devastates the once peaceful and happy New World home of so many millions of our countrymen. Our pianoforte makers might then also be expected to take to gun-stock contracts and the improvement of the Enfield rifle, our engineers to the building of engines of surpassing power and fitness, and our shipbuilders to the designing and construction of ships of war so truly formidable as to consign our *Warriors* and *Minotaurs* to the company of the harmless wooden tubs in the Medway, the Hamoaze, and Portsmouth Harbour. Resource, therefore, in the sense it is here spoken of, the genius to invent and the skill to fabricate what is invented, is an element of naval power second only in importance to that other and superior element of naval power which would strike a blow, and possibly a decisive one, before the enemy to be encountered had bethought himself what to do, or rubbed the scales from his doubting eyes.

Resource in Stores, Docks, and Seamen. Resource in stores, docks, and seamen are the only other elements that need be named under the second head, because with mechanical ingenuity and these all other things are in truth minor. All countries are more or less producers of naval stores, not meaning by the term the rosin, pitch, and turpentine of commerce, but naval stores in the largest sense, and it is a question that only extensive inquiry could solve, whether the country producing least, and continuing to produce least, is nevertheless practically as strong as the country producing most. Italy is the least store-producer of all the Powers. For lack of skilled mechanics and manufacturing establishments, it is obliged to build its ships of war in France, England, and America; but once all its now constructing ironclads are afloat, there is no reason to suppose that with a moderate supply of spare rigging, cordage, sail, and the like, these ironclads might not successfully open a great campaign. Stores to ships of war are, in the main, like suits of clothing in one's wardrobe, it being impossible to put everything on one's back at once, and there must indeed be hot work before two suits of ship's canvas and spare rigging, &c., are worn out, just as there would be something to show for the disappearance of a using and a reserve pair of trousers and the corresponding coats. Essential as stores are, they are, therefore, relatively of small account; and Italy without roperies, sail-lofts, and other old-fashioned ship-of-war adjuncts may be as powerful as Italy possessing them. Docks are indispensable, but they may be increased unduly, and before there is occasion for them. Ships of war do not always want to be in dock, any more than individuals at their hairdresser's. Ships of war usually require to go into dock to get their bottoms cleaned, with brush or scraper, and when cleaned and coated with Peacock and Buchan's or some other person's composition, a dock may not be wanted it may be for a year or more. Still, there are a class of people always dinning us about the number and capacity of the French docks, totally oblivious of the fact that the *Great Eastern* is likely to have to wait some time for a dock large enough to take her in, and that after one shall have been provided the owners may still be sufficiently self-willed to use a gridiron. No one ever cares to think that in this country there are at least a dozen times the number of docks that there are in France equally available for war or peace. And there is this to be said, too, as against France, that whereas the docks of Toulon, Brest, and Cherbourg admit of no enlargement, unless by great labour and expense, our docks—that is, our private docks—in the greater number of cases, may be made as long, as broad and deep as we wish, by a simple application of the spade. On the Isle of Dogs, on Southampton Water, and almost anywhere round our coasts, a few thousands of sturdy convicts would, in a week or two, excavate sufficient dock accommodation for the navies of the world, while any one of fifty large ironworks were preparing the gates or caissons. Certainly the best appropriation of useless space in our seven dockyards would be the construction of ornamental docks, which would be always ready, and the maintenance of which would furnish the minimum of excuse for staffs of tax-eating office-holders. Seamen are, of course, a great resource. Put them, however, on fast ships like the *Alabama*, and their peculiar qualities, even in a time of war, will not be often called into requisition. Hereafter, indeed, their part will always be subordinate. But after the ability to construct, equip, and fit for sea, on the shortest notice, ships that will surpass all others in speed and power, sailors follow, and while the best seamen will always render the best account, the greatest number will, in the end, be sure to prevail. Seamen, however, are not so much to be judged by numbers or by individual qualities, as by the fact of being free or not free. An inscription list of seamen is not worthy to be counted with half the number who are seamen by choice. Occasionally the inscription list of France may be mentioned to silence an inconvenient statement in the House of Commons; but France just now is, and always has been, distrustful of the seamen's services that are involuntary. Commission after commission has inquired into the subject, and if the truth were told, half the number of French seamen volunteers would at any time, in peace or war, be gladly taken for the aggregate that is returned. Nor is freedom alone the sole test to be applied to seamen.

A seaman to be really worth his salt must be intelligent. He must be a fellow that, as the saying is, "knows a thing or two." Among British navy seamen, ignorance may abound; but among French navy seamen, netted in as they are from stagnant and unknown fishing hamlets, there is vivacity without the self-reliance and independence that springs from the consciousness of being free, and, on shore at least, of enjoying freedom on the same footing and in the same measure as the best born and proudest in the land. What this consciousness and intelligence make the British sailor is only to be judged on the lower deck of a French or Russian ship of war.

<small>Immaterial whence resource is derived.</small> It is necessarily immaterial whence resource is derived. If a fleet have burned their coal, what matters it apart from expense whether the coaling takes place in the Tyne or the Tagus, at Portsmouth or Southampton? Coal is wanted, and when it has been supplied the coaling is completed, and the ships may proceed again to sea. If topmasts and other spars have been carried away in the North Sea, why not as well put into Copenhagen as make for Sheerness? Spars are wanted, and spars are much the same all the world over; with this difference, that the farther you go for them—namely, to Puget Sound—the better they are believed to be. If fresh provisions are required, and the fleet cruising in the North Atlantic, the Western Islands, and not one of the victualling yards, is the place to send to. "No, no!" exclaim the pensioners of the dockyards in all these cases; the Admiralty instructions are to be adhered to. They see no farther than their sordid interests will allow. That place and pension may be preserved, the resources of

the navy must be passed through the dockyard filter and their unworthy hands.

The third real element of naval power. Endurance. The third and last real element of naval power is endurance. This, perhaps, is the most popular dependence of all. Englishmen, it is believed, will not only fight better than all others, but they will endure more. In this belief there is a mixture of truth and untruth. Of the latter, because the present generation of English seamen have not been much tried, and because there is no enlightened desire abroad that a large number of them should be trained for ship-of-war service; the very basis of endurance in action, whether on sea or land, being perfect familiarity with gunnery and the use of small arms. Those, therefore, who take the superior endurance of British seamen as a matter of course, without caring whether they are trained or no, are merely deluding themselves and others. The British seaman who can only fight with his fists or with a marlinspike, however intelligent or patriotic, would have to yield without enduring anything before the untidy and semi-savage seamen of the Czar. Untrained seamen in action would be no better than untrained soldiers in the field. Endurance, therefore, is a questionable resource. That trained British seamen have not their equals on the ocean in all proper sailor qualities, is not for a moment to be doubted; but it is an illogical and false generalisation to step from the qualities of the few and assume them for the many. A Chinese Mandarin visiting the House of Commons might as reasonably infer from what he saw and heard that all Englishmen are affluent and well educated, while the truth is that very few are rich and many thousands are unlettered. Easy as it would be by inducements of a material kind to train 100,000 British seamen for reserve service in the fleet, until the thing is done we cannot venture to speak of British seamen as a whole able to endure what others would be sure to yield under. The mere endurance of wet clothes, bad or short supplies of food, is something quite different from facing and encountering cutlass cuts and revolvers, and with these all sailors in the service of their country in a time of war might any moment be required to grapple. At close quarters the partially trained inscription French seamen, although not a match for British seamen, would scatter raw British seamen from the mercantile marine in much the same fashion as chaff before a storm.

<small>Colonies neither a strength nor weakness.</small> Colonies, whatever may some time or other be made of them, are not at present a source of strength to any Power, and they can hardly in any case be deemed a weakness. In the event of war this country might seize the colonies of an enemy, but were peace afterwards to depend on the restoration, it is not conceivable that the right of conquest would now-a-days be insisted on. The spirit of the times is opposed to humiliations, and leans strongly the other way. Therefore it may be said that the possession of colonies will not weaken any Power. England occupying Algeria during a war with France would only too willingly surrender the unprofitable trust, and France occupying the Cape, Van Diemen's Land, or our Pacific possessions, would no doubt do the same. Colonies are valuable only as lucrative openings for a wandering but enterprising population, and as Frenchmen are not given to wandering, British colonies would be of little use to them; while, on the other hand, England has too

many colonies to be at all covetous of those of France. So the presumption is that in a war colonies, in respect to weakness, would count for nothing. As a rule they will be left to themselves, and should it not be worth their while to set a stray cruiser or even a strong squadron at defiance, the enemy no doubt will make allowance for their straitened circumstances, or take the benefit of their want of spirit. That colonies in many cases could be made serviceable to the mother country in extremity is true, but it is equally so that there are no end of obstacles in the way. Colonists, whether those of England, France, or Spain, possess and manifest all the impracticability of children, and from indulgent motives of State policy the parent nations spoil them. Take the case of Canada, which is as good as any, although there are many points of difference between it and other colonies, and especially between it and the colonies of France and Spain. Canada, without inconvenience to itself, could long ago have benefited the mother country to an incalculable extent by embodying 100,000 or 150,000 militia and volunteers. It will, however, do nothing of the kind, although the integrity of the empire might be regularly threatened in a semi-official way from Washington once a week. Why should Canada do such a thing, presents itself in a dozen different forms to the people of the province. One will ask what Canada has to do with the quarrels of this country. Another will say that Great Britain ought to defend its colonies. A third will say there is no danger. A fourth will say supporting the proposition would be supporting an unworthy Canadian Government. And so on. The consequence is that Canada does nothing, or so very little, that it is entirely useless. Meanwhile the home Government is powerless, and manifestly the part of wisdom under such circumstances is to take matters as easy as the colonists. Should Canada ever be invaded by the armies of the United States, and recovery appear a greater object to us, than the continuance of war would be an evil, all we would have to do would be to allow the war to proceed. Canada might contribute greatly to our material strength, but never will, unless the formation of a single regiment of unemployed enthusiasts is to be considered something. Australia and India might also aid us, but it would be as unsafe to count upon them as for Spain to cry to Cuba, or for France to appeal to Algiers.

<small>Probable immunity of Colonies in modern war.</small> On the whole it is probable that in modern war the colonies of belligerents will not suffer molestation. Even in the case of war between this country and the United States, were we to assume, on the part of the United States, a desire to annex Canada, there are many reasons why the United States to gain its object should conciliate the colonists by magnanimity, rather than estrange them by desolation and barbarity. Coaling stations, and places provided with docks and gridirons, would, however, be keenly fought for, but in such cases belligerents in consulting the interests of the people would only be caring for their own. Practical immunity would, therefore, be enjoyed from the scourge of war even in these likely cases. Modern war seems from necessity to impose this policy of consideration and generosity on belligerents, because its weapons, unlike those of former periods, will be more formidable and decisive by concentration than diffusion. In a war between France and England it admits of demonstration that it would be the height of folly for either Power to waste its energies in the trumpery

attacks or trumpery harassings of commerce which were so serviceable in bygone years. France would bring its whole available strength to bear on England, and England would bring its whole available strength to bear on France. A departure from this policy would almost inevitably involve discomfiture and disgrace. Even in a war between England and the United States, it would be rather a nice point for the law officers of the Crown whether the common law of Europe as regards privateers would not be held as applying to American cruisers; England hoisting privateers to the yard arm. According to the common law of Europe, privateering is and remains abolished. So is piracy. Obviously an affirmative decision is alone required on this point to require the United States to meet concentrated force with concentrated force. Stop privateering with a high hand, and Portland, Boston, New York, Philadelphia, and Annapolis will be covered by the whole naval power of the United States. Nay, say not a word about privateering, but transfer the naval strength of England to the shores of the United States, and who but a madman would counsel the frittering away of the naval strength of the United States in any form whatever, even in privateering? Thus it seems that colonies at least will escape. Doubt may rest on the decision of the law officers of the Crown, and on the intelligence of American Secretaries, but on the whole it is improbable that in a war between France and England the colonies of either would change hands, or be thought of by the belligerents; and almost equally so in the case of a war between England and the United States, it is improbable that Canada would be subjected to inconvenience or a change of masters.

NAVAL POWER. 183

Conditions of Privateering. If the conditions of privateering are calmly considered, various reasons will present themselves to show that the usage is no longer worthy to be counted among the elements of strength or weakness in modern war. Take the case of a war between the United States and England. The New York Chamber of Commerce passes a string of resolutions calling on the President to sweep British commerce from the ocean. Let us assume that the President is foolish enough to listen to the representations addressed to him and to give them effect. New York, Boston, and Philadelphia may enjoy all the privateering that they wish. What then? American sailing ships engaging in the business would sooner or later be swept from the sea themselves. But there are American steamers. True; but they are few in number, at least the sea-going portion of them. Let us assume that by an extraordinary effort a hundred sail put to sea, and an extraordinary effort would not succeed in equipping half the number. But the one hundred sail set out well armed and found, and each steamer has coal on board for a week's full steaming. Where are they to go to first to sweep British commerce from the ocean, and next to avoid British cruisers? Manifestly it will be wise to stay near home. A week's supply of coal is the full measure of their aggressive power, the full length of their robbing tether. When the coal fails, the steamers must be burned to avoid capture. France dare not, Spain will not, Holland must not, bestow a single sackful. No doubt the Confederate ships of war obtain coal and pursue their depredations; but were the United States as strong just now as England would be in a war with the United States, no Confederate ship of war would obtain enough coal to light a fire with. England, in a war with the United States, or in a war with France, would think of its interests and honour, and of its law books only when the others were fully cared for. Who thinks of the Sabbath day when his ox falls into a pit? True, during the Russian war Prussia was not chastised for its perfidy to the allies, but it was because the allies were not great sufferers by the disloyalty and wrongdoing. Had Prussia rendered half the service to the Russians that neutrals have to the Confederate States, Sir Charles Napier's orders and inclination and the temper of the English people were such that war would have been declared. Let us, therefore, think generously of the inconsistencies and extravagances of the Federal cruisers, and at the same time let the Government at Washington realise the fact of its relative weakness in a war with England.

Prizes and their crews. If, in a war between England and the United States, coaling is a barrier to successful privateering, prizes and their crews would almost be as great a difficulty. It would be no easy task to take captured ships to American ports, a hazardous experiment to burn captured ships, and very troublesome and dangerous to land captured crews on neutral territory. Were English ships to be burned by American privateers, there is no reason why English cruisers should not retaliate by burnings on the land. The burning of English ships would be the burning of private property, and the burning of American towns would be the burning of private property,—a game at which British ships of war would play with terrible effect. Thus, even were privateering to be unwisely sanctioned by the President and practised by American shipowners, it could scarcely be continued. Not one of the conditions of privateering with effect now exist. When ships derived their motive power from the wind there was practically no limit to the time that a privateer might keep the sea and effect captures, but no fighting ship is worth anything now-a-days without coal and speed. And when all ships of war were propelled by wind and canvas, it was practically impossible to maintain close blockades, so that openings could be always counted on in passing storms and continued gales. Now, a well-appointed fleet ought to seal a coast, the ships taking coaling turns, or all drawing their supplies from tender or transport. That the Federal ships of war have not sealed the Confederate coast is owing, first, to the great length of coast; next, to the want of steam transports; and, last of all, to the short supply of coal in the United States.

Chapter VI.

POSITION OF THE POWERS.

England's Dockyards.
In the present state of the dockyards there is manifestly great danger. The Department of Works, instead of being a department of frittering and jobbing, ought to be a department for keeping the dockyards in constant readiness for war. The director of works (no reflection being intended on the distinguished officer holding the position,—the wretched Admiralty system, and not he, being to blame,*) should be charged with the simple duties of providing for the docking and repair of disabled ships, the speedy coaling of ships on the active list,† and the reducing and maintaining of the dockyards in a state to suffer the least possible injury in the event of bombard-

---

* To prevent misapprehension, other Admiralty officers are throughout spoken of in the same sense.

†,Mr. Palmer, at the last meeting of the British Association, gave some account of iron shipbuilding in the Tyne, Wear, and Tees from 1842, when the first iron steamer —the *Prince Albert*—left the Walker slipway, to the present time. The competition of the railways in the carriage of coals gave the first great impetus to the building of screw colliers, and from 1852, when they first began to be employed in this trade, 5,212,713 tons of coal have been carried to London by them. Since its first introduction, too, the screw collier has been greatly improved, and the facilities for loading and discharging very largely augmented. The screw collier *James Dixon* frequently receives 1,200 tons of coal in four hours, makes her passage to London in 32 hours, there, by means of the hydraulic machinery which the President invented, among the other inventions which distinguish his name, discharges her cargo in 10 hours, returns in 32 hours, and thus completes her voyage in 76 hours. The *James Dixon* made 57 voyages to London in one year, and in that year delivered 62,842 tons of coal, and this with a crew of only 21 persons. To accomplish this work on the old system, with sailing colliers, would have required 16 ships and 144 hands to man them.

ment by an enemy. Just now the director of works is charged with none of these duties, nor is any other official in the service. He, like many others, is a mere Exchequer milkman. If he can squeeze out of the Treasury a few more thousands annually than the Treasury designed to give him, he, no doubt, believes that he has done something meritorious. If he can pile up buildings so as to heap up rubbish in greater quantities than are to be found in Toulon, Brest, and Cherbourg, he perhaps feels himself entitled to the Order of the Bath. Of the probable bombardment of the dockyards, and the desirability of there being little in them to burn, he never dreams. Of the necessity that exists to provide more adequately for the coaling of ships he has no notion, and very likely is as profoundly ignorant of what the Tyne and Thames have done for consumers, as of what France has done, and is still doing, for its navy. Of the utter insufficiency of the dockyards to provide on the instant for a disabled squadron, there are many reasons for believing that he is quite unconscious. The consequence would be, were we suddenly involved in a great naval war, that, like the victims of the garotters last winter, we would have to suffer. An action in the Channel, resulting in a drawn battle, might be followed by a blow from, say, the more speedily overhauled and coaled ships of France, which would consign the dockyards—the great national repositories of fireworks—to the flames.

The inexpensive means of placing the Dockyards on a war footing. The inexpensive means of placing the dockyards on a war footing not only stare us in the face, but are a source of great uneasiness at home and in the colonies. Turn the convicts into the dockyards, and while they will neither eat nor drink more than at present, nor require more clothing, they may be made to invert the existing state of things in a short space of time. Look at the convicts at work in Toulon Dockyard, and you will see in a moment what can be got out of forced labour. This poor fellow with the green cap is an Arab, and his crime was that of turning on the oppressor at the door of his own hut in Algiers. He cut the Frenchman down, and the judge who tried the case thought the circumstances were only such as to warrant him to sentence the poor Arab to the Bagne for life. His task from morn till night, and year to year, is to pull a heavy pontoon across the dock entrance between Darse Vauban and Darse Castigneau, as often as visitors, workmen, and other *forçats* desire to cross over. To his help mechanical ingenuity does not come, for there are no blocks with purchase and no winch to make the burden easy. The poor Arab has a chain such as that riveted on his limbs, and unaided he pulls the pontoon over. It is such another pontoon as that used for taking workmen and visitors to the Essex side of Bow Creek from the offices of the Thames Shipbuilding Company's yard, and with which, mechanically assisted, the old Irishman makes so much to do. You gaze mournfully at the Arab, cross over with him repeatedly, approach him backwards, looking at the *Invincible* ironclad in the roads, and learn the story of his wrongs. It may be unwise to do so, and still worse to give the circumstance this publicity. If so, let France not be so ungenerous as to blame the Arab. But to return. This Arab is doing the work of three or four men, and he does it well and willingly. Here are two Frenchmen *forçats*, and both in green caps too; for each has killed his man under circumstances which were thought to some extent extenuating. They are in the Bagne for life. Their ankle chains are riveted together, and, it is said, never to be unloosened night nor day until one or other dies. It is needless to say what their crime was; it was too horrible. One has been guilty of shaking his chain over night, he says involuntarily, because of the soreness of his ankle from a hurried step, the iron making sad havoc with the flesh. The offender has to do his mate's work and his own for a week by way of punishment. His task is to wheel round shot of old date and pattern to the foundry. It is trying work. He has been at it from morning notwithstanding, and is as energetic as if trying to win a wager. On the barrow he piles the shot without the intermission of a second, and when it is full he staggers off with it, dragging his chained mate behind him. So he continues throughout the day and throughout the morrow. No eye is on him; he has a task to do, and it must be performed within a given time. He therefore does it. Further on here are more green-caps who have killed their men, and who are engaged at excavation. Neither on them is there any eye, but each performs the work of two common navvies. Terror obliges the poor *forçats* to exert themselves as free men never do, although stimulated by unlimited earnings on task and job.

<small>What French *forçats* do English convicts may do.</small> What French *forçats* do English convicts may do. They may excavate docks and perform the general work of labourers. True, by association with workmen they may get snuff, tobacco, and other things, and learn something of the current news; but what of that? Why should they not get these, as well as other things? Military officers in charge of convict prisons will, of course, shudder; but

let them shudder. One has only to see the French *forçat* system in operation to arrive at the conclusion that Sir Joshua Jebb was a fool. France is very grateful for all extraneous aid in the maintenance of the *forçats*, come in whatever form it may. Step into the bazaar, in the Bagne, in Toulon Dockyard. Here are half a dozen respectable-looking *forçats* acting as salesmen, and as importunate as the charming creatures behind our own shop counters, and ten times more extortionate. This cocoa-nut shell, carved no doubt exquisitely, is 150 francs. This dressing-case, made from straw or bone, or a mixture of straw and bone, is 200 francs. You shake your head, and are offered a cross or a Virgin Mary made from a shank-bone, or a pin-cushion made from a cow's hoof, and the price of the two first is 40 francs each, and of the last, 20 francs. Something you must purchase; and while you make up your mind your official guide sits on a stool and gossips with the shopmen *forçats*, and sets you down as a shabby fellow unless you buy something handsome. Sir Joshua Jebb's system scorns this sort of thing, and rather than the convicts should do the State some service, and get tobacco, they are penned up like sheep or forced on the colonies. Turn them into the dockyards. There is a few years' work for them clearing away the rubbish that Parliamentary votes have accumulated to an extent which unfits the dockyards for war. After the dockyards are put to rights, why not require the convicts to construct refuge harbours on the coast? For all the crime of the country there is within the country ample work, if only the right sort of men were at the Admiralty, and the right sort of men were intrusted with the dockyard works.

England's Ordinaries. The insufficiency of the existing system of ordinaries is as great as could possibly be devised, although the design of the Admiralty really were to maintain a useless navy. Here is the mythical *Orion* frigate fresh from the West India station, where she was the crack ship of the Admiral. She is to be paid off into the third division of the reserve at Chatham. Out, therefore, come her moveables, and her masts and spars, which well might be deemed immoveables. She is now in a thoroughly useless state. Were we suddenly to become involved in a great war, the enemy might, in a dash at Chatham, destroy the *Orion* and all the other thoroughly useless ships at their moorings. Paying off the *Orion* into the third division would, in such a case, have been equivalent to paying off the ship to be burned. The ship, from a condition of high efficiency, was reduced to a condition of utter inefficiency, so much so that she could not be defended from attack. As the *Orion* is, so is the great bulk of the ships of war constituting the British navy. Our naval authorities, in some unaccountable way or other, have contracted the bad habit of playing at making and unmaking ships. They rig, that afterwards there may be the pleasure of unrigging. They are boys of the Marine Society, or of the late Mr. Green's school-ship, but boys of larger growth. When a British ship of war enters a foreign harbour, or approaches another ship of war of a friendly nation on the high seas, or in a roadstead, the orders are that the British ship shall be prepared for action so as instantly to resent aggression, should that occur. The rule is a wise one, seeing that it does not happen to be much abused. But why not extend its operation? Why not have the reserve ships of war in constant fighting trim? They would form unexceptionable barracks to our idle navy seamen, because the seamen would be always under more restraint than in barracks on the land. They would also enable the country, with comparatively little trouble and expense, to strike before an enemy was prepared.

<small>What the Ordinaries are.*</small> Dockyard ordinaries is merely another name for reserves of ships of war. When ships of war are said to be in ordinary, the meaning is that they are in one of three stages of readiness for commission and active service. In the first-class ordinary, ships are fully rigged, and in want of nothing but crews and powder to enable them to proceed to sea. In the second-class ordinary, the hull and machinery are supposed to be seaworthy, and the lower masts are in; but there are no stores on board. In the third-class ordinary, ships are in all conditions from the serviceable ship just returned from foreign service to the rotten tub that may have been at the same moorings for twenty years.

<small>Where established.</small> The ordinaries of the British navy are chiefly if not altogether established at Chatham, Portsmouth, and Devonport. This is an obvious, necessary, and unobjectionable arrangement, because were ordinaries maintained at Deptford or Woolwich, the navigation of the river would be impeded, and were they maintained at Sheerness or Pembroke, any stray cruiser

---

* The *Cochin*, gunboat, is from the steam ordinary, and has been brought into dock to be seen. She is in the first class; and as seamen, provisions, and powder are at hand, the *Cochin* might proceed to sea and engage an enemy in a couple of hours. Casting the eye for a moment in the direction of the gunboat ordinary, that little forest of small masts indicates gunboats in the same class as the *Cochin*, all of which might have pennants flying and men at quarters before the next train leaves for London. Hence the importance of reserves of trained seamen.—*The Dockyards and Shipyards of the Kingdom.*

might fire shells into them, or otherwise commit them to the flames. But in the Medway, in Portsmouth Harbour, and in the Hamoaze, the ordinaries are never likely to be disturbed unless by a squadron or a fleet. Batteries, guard and flag ships, interpose their guns between an enemy and the moveable reserve stock in trade of the British navy.

<small>Admiralty regulations respecting them.</small> The Admiralty, so prolific in rules, have, of course, framed many for the ordinaries.* As some of these regulations suggest impropriety and absurdity in as strong terms as could possibly be employed, it will be well to reproduce them. "When a ship is ordered to be paid off, the stores of every kind are to be returned to the department to which they severally belong." Distance, expense, and trouble are, of course, dirt-cheap commodities in the estimation of the Admiralty. "The rigging is to be bighted in regular lengths, and properly tallied before being landed—care being taken in clearing it that it be not chopped or otherwise injured." Of course, it is often chopped and injured, so as to be useless; but what does that matter to the Admiralty? "Lighters belonging to one department are on no account to be used for the conveyance of stores belonging to another department." Such a proceeding would, no doubt, be sufficiently awkward to oblige the Admiralty to discharge a number of the dockyard beef-eaters. "While under orders to pay off, more than ordinary precautions are requisite to prevent any peculation of the stores; additional sentries are to be posted at different parts of the ship, and the police directed to be very vigilant in observing

---

* *The Queen's Regulations and the Admiralty Instructions for the Government of Her Majesty's Naval Service.* London: Harrison and Sons, 1862. Price 1s. 9d.

that nothing be surreptitiously passed through the ports or out of the ship." As, however, love laughs at locksmiths, so once in five years, it is said, a dockyard town marine storedealer retires on a competency to a country box and drives a trap. "If, when a captain or other officer is appointed to command one of her Majesty's ships, any additions or further alterations are deemed advisable, application is to be made to the Admiralty, &c." So, as often as a ship is commissioned from the reserve, she is subject to such alterations and additions as the commanding officer may desire. When Sir Baldwin Walker, the late Controller of the Navy, commissioned the *Narcissus*, his own comfort and the comfort of his family required the construction of a poop. When the *Narcissus* is recommissioned, Sir Baldwin's successor may not like the poop and have it torn out. Think of one tenant getting another story to a house, and another tenant getting the story removed, and the absurdity of the system will at once appear.

<small>Ordinaries partake of the character of duplication.</small> Ordinaries partake of the character of duplication. They are a mere aggregation of duplicates, too often augmented with as little reason as that which justifies the usual practice of the dockyards in minor matters. The practice in the dockyards is not to be content with one of a sort, but to possess two. For example, in each dockyard there is a factory department and a dockyard department, the one being as near as possible a duplicate of the other, in respect to machinery and other things. Then again, for each ship of war there are duplicate boilers, either in store, in hand, or in contemplation. Such is the system of duplication. It represents the period when Portsmouth and Plymouth were long journeys from London,

and when it was thought desirable to provide against the contingency of a general break-down in one establishment or department. Precisely the same reasons that would justify the erection of duplicate engineering establishments with identical machinery within the same dockyard walls would justify the excavation of another set of London Docks, to be available in case the present set should by any chance become unserviceable. Precisely the same reasons would also justify the erection of new Admiralty Offices, and the formation of a new Board of Admiralty, to be in readiness in case anything should happen to the present Board of Admiralty or to the present Admiralty Offices. In a word, the occupant of a house might as well provide himself with another house, and the housekeeper provide himself with two legs of mutton instead of one, to guard against the possibility of one running off or being stolen. But it will be said there is this difference with regard to ships, that one dozen or even a score of a class may be inadequate. This is true; but in the absence of any rational rule, or even of any irrational rule, as to the accumulation of ships in reserve, it is equally true that the ordinaries partake of the senseless character of mere duplicates. Ships are multiplied as if for no other purpose than keeping the dockyards going. And yet, for the purpose of arriving at the proper strength of the fleet in time of peace, there is scarcely anything more required than to average the strength during the years immediately preceding the Russian war. This ought to furnish an unexceptional standard of the extent to which ship accumulation should be carried in a time of peace. And as regards war, surely an approximate estimate of waste might be formed from the experience of the Russian war; the important consideration being kept in view that the resources of the country are constantly available should an emergency arise. That is to say, were the estimate to prove inadequate for a lengthened war to which reserve ship accumulation might be carried, there would be no difficulty in adding to the reserve as occasion might require—to the extent almost, if not entirely, of one hundred efficient ironclads per annum.

<small>Objections to the system. Waste.</small> To ordinaries, therefore, a variety of objections may be urged. That of waste involved in returning stores has been already noticed. But there is a broad view of the question. The ordinaries represent the product of the sums voted annually by Parliament, and these sums of late years have been on the extraordinary scale of £250,000 a week, £1,000,000 a month, and £12,000,000 a year. Legitimately and wisely used, the expenditure of a single year should suffice to enable us to suspend all ship of war construction for a lengthened period, because at £50 a ton there is a most gigantic fleet of ironclads in so much money. But the expenditure practically yields us nothing. True, we have the noblest ironclad fleet afloat, but the construction has been spread over several years, and the existence of the ironclads makes no perceptible addition to the reserves. We are getting ironclads, and room is made for them in the reserves by weeding out. On the one hand we increase, on the other hand we decrease—by waste. We get no nearer that which should be the end of shipbuilding, at least in a time of peace—stopping. There is an obvious screw loose. Naval expenditure does not carry us, as it ought to carry us, to some cumulative result. The Admiralty are a sieve through which the hard cash voted by Par-

liament runs like water, and when the money is all spent towards the close of the financial year the only sensible effect is the wetting of the wires. Lord Clarence Paget won his spurs in the House of Commons by declaiming against this waste; office has since opened his eyes—or closed them.

<small>Our neighbours' system.</small> In France they manage these things better. Visit Toulon, or any of the other French dockyards, and there is none of the unsightly dilapidation characteristic of the British ordinaries. The French give their ships a turn in commission, but those out of turn are neither out of sight nor out of mind. On the contrary, they are always cared for and seen to. Frequent surveys are held, decay is checked, defects made good, improvements introduced, and general sea-going readiness constantly maintained. Were an enemy appearing off Toulon, or were war declared, most of the reserve ships might proceed to sea within four-and-twenty hours. The riggers have merely to go on board, while the convicts deliver the stores and the sailors stow them. The French system is, therefore, the complete antithesis of ours. The advanced French ships are in the roads, the second class have their masts in and a portion of their crews on board, and the third class are ready for their masts and other things. The insane gutting of paid-off ships is not practised, but France economises its ships as it does its regiments. French ships in ordinary resemble French troops idling in garrison towns, who only require food and powder to enable them to enter on a campaign. English ships in ordinary, on the contrary, resemble the broken-up and broken-down veterans in Chelsea or Greenwich Hospital, who want almost to be born again to become effective.

France, of course, has learned the secret of spending its money properly, and we have that lesson still to learn.

England's Shipyards and Workshops. If it is the case that the condition of the ordinaries and the dockyards is bad, there is some comfort in the reflection that England excels all countries individually and collectively in the material and appliances of modern shipbuilding and war. The ground on which we tread abounds in coal and iron; and capital, the aggregate amount of which perhaps baffles accurate computation, is set apart and devoted to their appropriation. Belgium also possesses coal and iron; Sweden is bountifully supplied with the latter; and America has coal and iron fields of unknown area; but practically it is in England only that the yield of coal and iron assumes proportions at all commensurate with inexhaustible supply. England is the coal and iron market of the world. Naturally, in a country so happily circumstanced, teeming with particular forms of mineral wealth, and with capital ebbing and flowing in constant and copious streams for its realisation, every branch of mineral industry is developed in an extraordinary degree. Its iron workshops and iron shipyards are unrivalled. Of those of Belgium, America, and France, much may be often said, but it is because Englishmen have not yet learned how far in iron products they have outstripped the world. If in Belgium there are a few great foundries, the number is many here; if, in America, Baltimore, Pittsburgh, and St. Louis are the seats of extensive armour rolling, the English rolling-mills will yield a foot or more for every American inch; and if in France the Imperial Government have great engine and repairing works scattered round the coast, there is not one of them on so large a scale, so well stocked with machinery, or so productive, as the great private engine shops of this country. France has no such establishment as that of Messrs. Maudslay, Sons, and Field, at Lambeth, or as that of Messrs. Penn and Sons, at Greenwich. France has no such collection of engine and repairing shops as are to be seen on the Thames, the Mersey, the Clyde, or the Tyne; in fact, the engine and repairing shops of either of those seats of marine industry equal, if they do not surpass, the engine and repairing shops of the whole of France. So it is with shipyards. Nothing can be more absurd than the alarming comparisons sometimes instituted between France and this country in respect to shipyards. At this very moment there are several steamers building in this country for French companies, because they can be speedier and better put together here than in France; and if any one will take the trouble to glance through the remaining chapters of this volume, it will be found that it is not a new thing for English shipbuilders and engineers to supply the wants of the navy of our nearest neighbour.

Fallacy to be guarded against. From this great productive power the inference is not warranted that for immediate war we are stronger than other nations. On the contrary, it has been already shown that the foremost element of naval power now-a-days is readiness to strike a blow the moment war shall have been declared. England may surpass the world in ironworks of various kinds, and still be very weak from the circumstance of not organising its resources and rendering them available. As long as the ironworks of this country are unavailable for war, this country, in respect to such works, is in the identical position of the poor of St.

Giles's when passing well-filled bakers' shops. There the food is, but so far as the poor are concerned there might as well be none at all. To them it is unavailable. Let it, therefore, not be hastily deduced that because we possess vast mineral resources, and that capital to an incalculable extent is invested in numberless ways in the working up of iron, we are for those reasons in a most formidable condition. We are no such thing. As matters at present stand, before the iron industry of the country could be fully set in motion either for the purpose of defence or aggression, the country might be invaded, or its prestige sacrificed abroad. As a whole, then, the position of England in a naval sense is not such as any thoughtful Englishman can approve. Our dockyards are in disorder and unready, our ordinaries are inefficient if not absolutely in decay, and not more than a beginning has yet been made in rendering available the great resources of private enterprise. What private enterprise is capable of accomplishing in this country forms the subject-matter of succeeding chapters.

French Dockyards.[*] The French dockyards are most formidable. Defence is, and always has been, the fundamental theory of the French dockyards. To British resource and valour this is chiefly owing. Enterprising and heroic as France has been, both often, if not always, were unavailing when confronted with what is still the greatest maritime population of the Old World or the New. The only course, therefore, for France to pursue, was to make good against attack the great national establishments on which the naval strength of the monarchy, republic, or empire had been wholly reared.

---

[*] This view of the French dockyards I have before expressed in a popular weekly newspaper—the *Illustrated London News*, October 3, 1863.

To this necessity the engineering reputation of Vauban and the First Napoleon are largely owing. In this country, on the other hand, the time-honoured theory of the dockyards is attack, and it is because of this that military engineering has attained comparatively little prominence among ourselves. During the long war of the last and the present century, Spithead and our other great dockyards may be said to have had no defence but our wooden walls, and had engineering defence been resorted to, it would, not in England merely, but in France and Europe, have been thought an acknowledgment of exhaustion or decline. These different theories are to the present day the key to the grouping of our own dockyard factories, shops, and stores, and to the dispersion of those of France. We have studied mere convenience, never dreaming of defence or burning by attack; while France, with the certainty of attack constantly before it, has sacrificed convenience that the consequences of bombardment might always be reduced to the lowest point. The circumstance is one of great importance, because if now-a-days our dockyards are to be defended like those of France, it seems to follow that the first thing to be done is to reconstruct our dockyards by separating all the separable parts as much as possible from each other. Toulon Dockyard exemplifies the French dockyard theory. Instead of the Toulon establishment being on one side of the harbour, as it might be, it is on two, and the arrangement involves exactly the same superfluous labour as though one-half of our Portsmouth establishment were on the Isle of Wight. Within the five Mourillon ironclad building slips there are at the present time in hand three frigates and one transport, and while the timber entering into the construction is beside the workmen, the iron has to be carried backwards and forwards, either from the Maritime Arsenal or from Castigneau Arsenal. The inconvenience must necessarily be great. But what of that? for were Toulon attacked by an enemy, the burning of Mourillon would leave Castigneau and the Maritime Arsenal unharmed; and the converse. Nay, so arranged are the parts of Toulon Dockyard—unquestionably the best dockyard in France —that were Castigneau to be burned, the Maritime Arsenal would probably escape, and were the Maritime Arsenal to be burned, Castigneau would probably escape. But this is not all. Mourillon is, as far as possible, the repository of all the combustibles within the dockyard; so that let an enemy bombard as he may, no more, practically, than one portion of the dockyard can be burned. In our dockyards, on the contrary, timber is stacked everywhere, and roperies, sail-lofts, mast-houses, and general storehouses have been erected, under the sanction of the Director of Works, without the remotest reference to bombardment, in situations where at the time there happened to be room for them. Some of the roperies and other buildings are so situated that a single live shell falling in them would endanger the entire dockyard. Thus the French dockyards are formidable, while ours are weak. While ours would burn, theirs would hold out, offering, of course, all the time serious resistance from the forts and batteries, which in our case would be silenced in the destruction of the dockyards.

<small>French strength and weakness.</small> If France is strong in its dockyards, it is also strong in the advanced state of preparation maintained in its steam reserves; but there is a weakness in France that appears to escape general

notice. France has its weak points as well as other nations, with this difference—that those of France are palpably on the increase. Up to a recent period France was a most compact country, possessing few colonies, and presenting an unattractive and unimportant seaboard. It was then a country bestowing a few luxuries on other nations, which in the main were paid in coin; supplying its domestic wants by its own domestic industry. This obviously was a strong position for a Power to occupy. Times have since changed. France is no longer the exclusive country that it was five or ten years ago. Among its people a demand for something else than coin has arisen in exchange for silks and wine, olive oil, and other things. The raw and manufactured products of other countries are now preferred. French industry and commerce are at this moment receiving an impetus and expansion that have to be witnessed to be believed. Not merely its great interior seats of industry, but its great seaport towns, share equally in the change. For example, the Marseilles of to-day is not the Marseilles of only two or three years ago. It is extending in all directions, and a noble business town it is. It is the Liverpool and New York of France, and not in the least inferior to either in its business aspects. Great docks have been constructed, and are densely crowded with the noblest steam and sailing ships afloat. New docks, of as great if not greater magnitude, are in progress or contemplation, and the increasing trade of the port calls loudly for the facilities that the new works will afford. Most marvellous of all, the Suez Canal, which in London is a dream and an impracticability, is in Marseilles a reality. Not a sober-minded merchant doubts of it, and not a cautious banker hesitates to reckon it among the assets of his customers. And there is no speculative feeling on the subject. No one is speculating in probabilities in relation to it, because there is but one opinion on the subject, and there are none building castles in the air. All that one hears is that when the Suez Canal is opened, Marseilles will become the commercial emporium of the world. It will be the half-way house between England and the East, and the London of Continental Europe, as regards all Eastern products. Calmly considered, and from a purely business point of view, the expectation is wise and necessary. So Marseilles and the South of France prepare for the opening of the Suez Canal. Along the shores of the Mediterranean, the tall chimney-stalks of soap, sulphur, vitriol, leather, sugar, and other works spring up, infusing new ideas into the mind of every one. Along the docks of Marseilles, the first section of the Suez Canal dock warehouses has been built and finished. The second section is on the point of being begun, and when finished both will present a front more than a mile in length of storehouse palaces. The new warehouses of the Victoria Dock Company near Fenchurch-street Station are insignificant in comparison with them, not in one respect, but in all respects. Those who planned the Suez Canal dock warehouses at Marseilles did so with the intention of surpassing everything of the kind with which commercial men are familiar. All the new improvements of the Victoria Dock warehouses are embodied in those of Marseilles, with others of French and American origin, and the result is buildings such as the world has not yet seen. London is immensely outstripped in warehouses by Marseilles, and Liverpool and New York are incredibly behind. Obviously the France of such things is not the France that it was. The France of the Empire is pre-eminently become, or at least is rapidly becoming, the France of peace. All the advantages of isolation for the purpose of defensive war are being surrendered, and France becomes as much exposed to the assaults of naval Powers as Holland or the United States. It is impossible to overrate the importance of the change; and Englishmen, in helping forward the Suez Canal, and embarking more and more largely in French commerce, will be binding over France in solid and enduring bonds to keep the peace.

<small>Algeria.</small> Algeria is another new element of change in the defensive state of France. Twice a week steamers of the Messageries Impériales leave Marseilles for Algiers, and perform the voyage in less than forty hours. In a shorter time than is occupied in the voyage from London to Edinburgh, Dundee, or Aberdeen, the voyage to Africa is accomplished. In less than half the time consumed in the voyage from London to any part of Ireland, France communicates with its great and rapidly improving colony. Practically, therefore, for the purpose of defence Algeria is as much a part of France as Scotland or Ireland of England, although the former enjoys the advantage of the railway. And what is the state of things in Algeria? In Algeria the population is increasing rapidly and commerce expanding. Probably at the present time the intercourse between France and Algeria is of considerably more importance than that which existed between London and Scotland or Ireland five-and-twenty years ago. In this trade numberless Frenchmen are deeply interested, and the amount of capital at stake is necessarily very large. To this trade France must cling in war. It must defend it with fleets, squadrons, and cruisers, because it is worth defending, and because it is at the door. Were

Algeria in the North or South Atlantic, the case would be different, but as it is France has no choice. It must protect Algeria, although the remote possessions of England would in all likelihood be left to protect themselves. It thus behoves France to see to the efficiency of its navy, for its ships now-a-days have their work before them. So, it appears, also thinks the Minister of Marine, because a number of coast defence ironclads have just been ordered. Altogether the present position of France is assuring to all Englishmen. France is really no longer a menace to England, and the talk of the Mediterranean being a French lake is nonsense. England is infinitely more a menace to France than the contrary, and were England to be thoroughly prepared for war, long years of peace would in all likelihood be enjoyed. On that legislative course which England has long entered with incalculable benefit to itself, France has entered, wisely reckless of the consequences to which its relatively great exposure lays it open.

<small>The French Shipyards.</small> The condition of the French shipyards may be pretty accurately inferred from the state of the La Seyne shipyard, at the head of Toulon Harbour—one of the greatest if not the greatest shipyard in France. The building frontage and launching facilities are extraordinary, and the yard looks as roomy as an English dockyard. The roominess, however, arises from a striking cause: the iron shipbuilding machinery, the punching, slotting, boring, and other things are not on the spot, but in the engine shop of the company, near Marseilles. In the engine shop all the iron is prepared and shipped round the coast to the shipyard, so that the shipyard remains an open space and the iron shipwrights are the mere fitters of what is sent to them. The only

visible piece of machinery is a double punching and clipping machine, such as may be met with in a third or fourth rate English iron shipyard. Why this is so is that iron shipbuilding has yet made comparatively little progress in France. The machinery of the French engine shops as yet performs the work of the French iron shipyards. True, there are exceptions at Bordeaux and one or two other places, but the rule is as stated. France is behind England in iron shipbuilding many years. If war is to be waged with iron ships, France, therefore, must be a very peaceful neighbour for a long time to come. It is really no exaggeration to affirm that England is capable of turning out one hundred iron ships of any class for every French one. The case of the La Seyne shipyard is analogous to the case of a firm possessing an engine shop at Southampton and a shipyard on the Thames, all the iron being prepared at the former and sent to the latter. Such a firm could not maintain itself in England. The only advantages that France presents in the construction of iron ships are an extremely low rate of wages and extreme, because forced, pliancy on the part of the mechanics. Against these are to be placed the fine, compact establishments of this country, the nearness and abundance of coal and iron, and, last but not least, the honest independence of the workmen.

America the dangerous rival of England. America, and not France, is the dangerous rival of England. America abounds in shipyards, and if at the moment it has comparatively little coal and iron available, there is practically no limit to the resources of its mines. If also at the moment Baltimore, Pittsburgh, and St. Louis are behind the great private establishments of this country in forgings and armour-plate rollings, it is not to be doubted that a prolongation of the war will soon lead to the erection of works rivalling our own. And just now America has in reality a very formidable ironclad fleet afloat. American naval architects have not from the first doubted that even the Monitors are available for operations on our own coasts. On the low sides of these ships they propose to raise temporary iron topsides and a temporary iron deck for the Atlantic voyage; and it is worthy of remark that Captain Coles has had in contemplation the fitting of a Monitor in the same manner for Australia—that is, to make the Australian voyage. Before long, therefore, America would be in a condition to attack the Mersey and the Clyde with a fleet of such ships, and as they would find shelter where our long-legged ironclad ships could not follow them, it is impossible to think of the situation without a shudder. To this danger the public must become alive, and give over dreaming of injury from France. With ironclad ships of war of small tonnage this country would effectually overawe America, because such ships could ascend the St. Lawrence river and the Lakes, descend into Lake Champlain, and even threaten the Upper Mississippi across the Illinois Canal. But with large ships only we are really at the mercy of America. America may in such a case attack us, while comparatively little injury could be inflicted in return. With an efficient fleet of small craft we could lay waste the whole region of the Lakes and the Mississippi, while our large ships inflicted all the needful chastisement on the Atlantic towns. Finally, we need not disguise from ourselves the fact that the more violent American politicians look forward confidently to successful tampering with the Irish.

Chapter VII.

THE THAMES IRONWORKS AND SHIPBUILDING COMPANY (LIMITED), BLACKWALL.

Annual capability. These are most important works. The ordinary construction capabilities of the company are 25,000 tons of ironclad ships and 10,000 tons of first-class merchant steamers, all in hand, and progressing simultaneously. Pushed and working extra hours, this large amount of tonnage might be raised to 40,000 tons in the aggregate, and the rule is that one-half of the whole tonnage in hand is turned out finished every year, whatever the class of ship may be. In other words, in a time of pressure, the Thames Ironworks and Shipbuilding Company could be relied on for 20,000 tons of ironclad ships annually, which, at £50 a ton, represents an exact product of £1,000,000.*

Locality and plan of the works. Middlesex. The Middlesex portion of the works is immediately in the rear of Trinity-wharf and the shipyard of the Messrs. Green, Orchard-street constituting the boundary between the three. This portion embraces the offices, joiners' shop, plumbers' shop, and timber-yard. The stock of timber being always small, a row of gunboats or sloops might at any time be laid down, and finished, subject to the

---

* If this is thought an excessive estimate, then 40,000 tons of ironclad ships in two years and a half is really less than might be done.

disadvantage of floating the iron across the creek which separates the comparatively small Middlesex portion of the works from the Essex side. Until this disadvantage can be allowed for by the Admiralty in a time of pressure, or by the great shipping firms in brisk times, the timber-yard must remain as it is, so close are the calculations and keen the competition in modern shipbuilding. This is a noteworthy reflection. There is no doubt as to what this or that will cost, although it is impossible to say what the eventual cost of an ironclad will be, or the length of time over which the construction will be spread, if there is a change of plan once a week or month. Some half-dozen gentlemen meet once a week, and in an hour dispose of the business of this great company, just as half a dozen gentlemen will dispose of the business of a railway, making the investment pay. What a pity that the affairs of the Navy are not managed in the same manner; a Thursday forenoon board meeting, and such a man as Captain John Ford, the managing director of the Thames Ironworks, rendering an account of his week's stewardship to his colleagues. That such an arrangement is practicable is proved by the success of the Thames Ironworks, and by the management of our great railway companies. The Middlesex area of the yard is five acres.

The Essex side. Bow-creek, a conveniently broad and deep arm of the River Thames, separates the minor from the major portion of the Thames works. Into this creek the ships built on the upper slips are launched, while those built on the lower slips launch into the Thames. The farthest down of these last-named slips is 370 feet long, and the two others 400 feet; on the larger slips, ships of 12,000 tons may be built and launched with ease and safety. Higher up there are five more building slips, on the longest of which is her Majesty's frigate *Minotaur*. The first of these upper slips is 345 feet in length, the second 400 feet, the third 320 feet, the fourth 240 feet, and the fifth 314 feet. Between the fourth and fifth slip there is a dock inlet or creek, which might at any time be enlarged into a dry dock or basin for ships of the largest class. To the right of the farthest down slip stands the mould loft, and along the heads of the lower slips stand the angle iron binding shop, and some press sheds. Between the lower and the upper slips the great workshops are situated. These comprise the foundry of rolled and hammered iron, a range of smithies, a copper-foundry, and an iron rolling-mills shop; behind which are the forge and forge shop, the tool shop, and another press shed. To the rear again of these workshops there are the engineers' shop, the boiler-makers' shop, the angle iron shop, and the pattern shop range of buildings. In the extreme rear and skirting the first bend of Bow-creek are the saw-pits, saw-mills, joiners' shops, and foundries. Compactness with completeness has been the aim of the managing director, at least such appears to have been the case. Round the building slips, but at convenient distances to meet the requirements of modern shipbuilding, the various workshops are grouped, so that a step may only separate the workshops from the slips, and the maximum of result be obtained from skilled and highly stimulated contract labour. From the anvil to the slip, from the slip to the engineers' shop, and from the engineers' shop to the saw-pit and the joiners' shop occupies an instant, and dispenses with the services of what otherwise would prove an army of working labourers. The advantage of this arrangement is shared mutually by employer and employed. The workmen look at the task that is before them and at the means at their disposal for its accomplishment, and other things being equal, in comparison with a badly planned yard, the Thames Shipbuilding Company would receive more labour for their money, and the Thames Shipbuilding Company workmen more money for their labour. This is an economic principle of admitted soundness, and still one that is too generally overlooked. Employers of shipyard labour too often think that if they lay down slips and erect shops where it is convenient, or where a fine effect will be produced, their part is done; in other words, their best is done for their own interest: and, of course, few think of any interest that workmen have in their yards beyond the weekly dole on Saturday. But workmen have so deep an interest in the judicious planning of a shipyard that the time may come when all the best workmen will be found in the best planned shipyards, because in these their contract labour will and ought to be more productive than in shipyards where slips and workshops are far apart. And shipbuilders who look no further nor deeper than great effect are doing themselves unconscious injury. They may fancy it the workman's matter to walk forty or fifty yards when four or five should serve, but in reality it is their own. Diminish the obstacles of the workmen in respect to distance within the shipyards, and the result is equivalent to putting a new and more effective tool into his hands, the use of which hastens his operations, and, if it does nothing more, clears the building slips for new ships at an earlier date. Long distance, scattered shipyards are, in a word, laid out on the effete dockyard model, and close-packed shipyards are in harmony with the enlightened economic teachings that now prevail. The proprietors of the

Thames Iron and Shipbuilding Works may congratulate themselves on the sound practical judgment that directs their great interests. The Essex area of the yard is twenty acres.

<small>Railway and other facilities of the works.</small> The railway and other facilities of the works are very great. Passing by the facility of access from the City with the Blackwall Railway, the works being in sight from Brunswick-pier, the Victoria Dock and North Woolwich Railway skirt the forge and smithies at the back and communicate with them by branch lines, thereby bringing the works within the cheap and expeditious coal-carrying resources of the Eastern Counties Railway; and the Eastern Counties Railway being thus in communication with the works, the whole country becomes an available market for the great forgings of the steam-hammers. The hammers and rolling-mills of the works in the course of a year would supply shafts, girders, and other things to an extent that would be hardly credible, and the former, if applied to the manufacture of ordnance, would at any time, but particularly in a time of pressure, render great national service. With the Thames to launch ships into, it is needless to observe that the great engine-building resources of Messrs. Maudslay, Sons, and Field, and of Messrs. Penn and Sons and the other firms, are at hand.

<small>No such facilities possessed by the Dockyards.</small> That no such facilities are possessed by the dockyards is well known to those acquainted with the dockyards. Deptford, Woolwich, Chatham, and Sheerness doubtless connect with all the railways, but it is at a disadvantage that practically is prohibitive. Chatham and Sheerness particularly are so

much out of the way of the midland coal and iron fields that it is an act of sheer economic waste to establish within them the elements of great iron manufactures. Under the best working system that could possibly be devised, it would be utterly impossible for the Admiralty to compete with the great iron manufacturing establishments a hundred miles or more nearer the coal and iron districts. Then what shall be said of Portsmouth, and Devonport, and Keyham? They are even more out of the way than Chatham. They are at the one extremity of England, and without much exaggeration the coal and iron fields of England are at the other. Such being the case, who can justify the enlargement of Portsmouth Dockyard and the extension of Keyham, that step by step great ironworks may be erected, and the Admiralty, beginning with mere purchases or reclamations of land, shall end with foundries, armour forges and armour mills, and engine-building shops, rivalling those of the Messrs. Maudslay and the Messrs. Penn? But, returning to the facilities of the Thames Iron and Shipbuilding Works, and the inferiority of those of the dockyards: not only do the Thames Iron and Shipbuilding Works possess the relative advantage of nearness to the coal and iron fields, but where is the dockyard that is laid out in a manner so well calculated to construct ships of war cheaply and in a short time? There is no such dockyard in England, nor is there one in France. In Europe there is no such dockyard, and it is doubtful if there is one in America. The Thames Iron and Shipbuilding Works are contrived to economise time and labour, and yield a certain return on the capital of the proprietors, while the dockyards of England and Europe are in the main contrived for show. In plain English, and speaking from personal observation, the one is a utility and the others are impostures. He who asserts the contrary, and alleges that the dockyards are superior or even on a level with the Thames Iron and Shipbuilding Works for the construction of ships of war, asserts either an impudent or an ignorant untruth.

The capabilities in excess of those of all the Dockyards. The capabilities of the Thames Iron and Shipbuilding Works for the construction of ships of war are in excess of those of all the dockyards. As the case stands at present, it is that whereas the Thames Iron and Shipbuilding Company can build ironclad ships of war out of scrap and other iron worked up into angle iron, plates, and armour-plates, with their own fires and machinery, the dockyards, one and all of them, go into the iron market for all their angle iron, plates, and armour-plates, because they are not possessed of the means of producing either. It is therefore clear that the capabilities of the one surpass those of all the others. Not long ago, before the launch of the *Royal Oak* and the other converted ironclads, the Thames Iron and Shipbuilding Company had in hand tonnage little short of all the new and converted ironclad tonnage in all the dockyards; the dockyard new tonnage being, in fact, not more than the usual aggregate of an ordinary second-rate merchant shipyard, the ships being the *Achilles* at Chatham, the *Enterprise* and *Favourite* at Deptford, and the *Research* at Pembroke. The *Lord Warden* and the *Bellerophon* cannot even yet be added to the dockyard list. Thus, again, at ironclad ship construction the dockyards are a sham, which the Admiralty themselves confess by the attempt at dockyard extension, with which Parliament will shortly have to deal.

Peculiarity of the work. There is a peculiarity in the work of the Thames Iron and Shipbuilding Company to which attention requires to be called. The company are manufacturers of iron as well as shipbuilders. They make the material of ironclad ships, and afterwards make the ships. This is not usual, but is a peculiarity of the Thames Company and of one other company. All the other shipbuilders are in the same position as the Admiralty in this respect. They go into the market for what they want, and should the iron not be forthcoming at the time agreed on they must wait. Not so the Thames Iron and Shipbuilding Company. They require to wait for no one. On their own forges and rolling-mills they at all times can rely. This is a consideration of great importance, because the company the makers of their own iron are not in the hands of those whose interest it might sometimes be to supply an inferior article. No doubt it is possible for ironmasters to supply other shipbuilders with as good iron as any that the Thames Company can produce, but the point to be borne in view here is that this company with another are not dependent on casual supply. What they contract for they can accomplish, while it is well known that the most serious delays have frequently arisen in the ordinary shipyards from want of punctuality on the part of the ironmasters. Nor are the shipyards the only victims of such irregularity. The other day the *Observer* charged the contractors for the ironwork of the Charing-cross Terminus with hindering the opening of the line, certain girders and other things not being forthcoming at the time agreed. Within the Thames Company's yard all the ironwork of an ironclad is not only put together, but actually produced.

The works chiefly designed for shipbuilding. The *spécialité* of the Thames Iron and Shipbuilding Works is the building of ships. The company built the *Warrior*, the first of our ironclads. Since the construction of that noble ship the company have constantly had ships of war in hand, and are therefore the most experienced ironclad shipbuilders of this or any other country. Recently, at one and the same time, the company had in hand, and still have, with the exception of the Russian battery *Pervenetz*, which has just been safely delivered at Cronstadt, the following ironclads:—First, the *Minotaur*, one of the largest and most powerful iron-cased frigates yet designed for the British Navy. The length of the ship is 400 feet between perpendiculars; breadth 59 feet 4 inches, and tonnage 6,812. This ship will be protected from stem to stern with armour-plates $5\frac{1}{2}$ inches thick, resting on a backing of 9 inches of teak; the armour and teak being somewhat diminished in thickness at the extremities. Between the *Minotaur* and the *Warrior* there are some important points of difference. The length of the *Minotaur* exceeds that of the *Warrior* by 20 feet, and the breadth by 1 foot 4 inches, and only about three-fifths of the side of the *Warrior* is protected by armour-plates. Again, while the *Minotaur* will carry 1,800 tons of armour-plates, the *Warrior* carries no more than 900 tons. The total weight of the *Minotaur*, with engines, coal, rigging, and armament, will exceed 10,000 tons. Second, the *Valiant*, another iron-cased frigate of 4,100 tons, is being finished by the company for the British Navy in the Admiralty yard at Millwall. Third, the *Pervenetz*, iron-cased Russian battery of 2,800 tons. Fourth, the *Victoria*, iron-cased frigate of 4,860 tons for the Spanish navy; and fifth, another ironclad frigate of 4,300 tons for the Turkish navy. In addition to these ships of war, which alone would constitute a formidable navy, being indeed a force more powerful than the fine vessels now forming our Channel fleet, and superior to what even France at the present moment could send to sea, there is a large amount of merchant tonnage of the highest class in hand. These comprise—first, a screw steamer of 2,100 tons, for the Peninsular and Oriental Company; second, a paddle-steamer of 2,000 tons, for the same company; third, a paddle-steamer of 3,900 tons, for the French Compagnie Transatlantique, intended for the mail service between France and America. There have also just been vacated two slips, each suitable for building large vessels, and they are likely to be soon occupied by two paddle-steamers of 1,200 tons each, for which the contracts are about to be concluded. Another such company so occupied with war and merchant tonnage is not to be met with the world over. The contract for the stately *Minotaur* was taken in September, 1861, and the launch of the ship is fixed for November. Chatham, to be sure, has its *Achilles*, but it has no more; and the *Achilles* is in truth the only ironclad that all the dockyards have in hand; while the other day, by the side of the *Minotaur*, there were four ironclad frigates, besides the merchant steamships just named: the work on all proceeding simultaneously.*

The working system. The working system of the Thames Company is contract between owner and shipbuilder, and contract between shipbuilder and workmen. The company contract for ships at a certain price, and the

---

* Let it be repeated, the ordinary constructive capabilities of the company are, 25,000 tons of ironclad ships, and 10,000 tons of first-class merchant steamers, while in a time of pressure 40,000 tons might be proceeded with.

work is then let out to the workmen at another certain price; the difference constituting the fund from which material is replaced, interest on works allowed, expenses paid, and returns made to the shareholders. That the practice is a satisfactory one to all concerned stands in need of no further proof than the fact that within the past two or three years the company's works have doubled in area and resource. The Thames Company's works of to-day are the prosperous result of a comparatively small beginning. This is a fact on which the Admiralty might reflect with profit. The motto of the Thames Company is "no work no pay," and that is the motto of successful private enterprise throughout the world. In the Thames Company's works there is no superannuation, and there are no incompetents. Nor are the terms of hiring labour in a perpetual state of fluctuation. On the simple honest principle of giving men all the work that they can do, and paying them for all that they do, the whole working system rests. The other day, when the Russian battery left partially unfinished, no difficulty arose between employer and employed, or between the company and the Russian Government. So much work had been done, so much remained undone, and an adjustment between all parties perhaps occupied half an hour. Those who think that ships of war cannot possibly be repaired by contract should take a note of this. In the way in which employer and employed work together in the Thames Company's works, there may be a suspension of work at any moment on any particular ship without the interest of either suffering. The case of the *Valiant* is also in point. The Thames Company contracted for the finishing of the *Valiant*, and the ship is finishing in the usual contract manner. With the Thames Company the Admiralty may order any change whatever in their ships of war, and the charge takes the one form of cost as between the company and their workmen. The workmen know their business, and the many excellent heads of departments, under the direction of Captain Ford, know their business. Among all classes of the employed the utmost goodwill prevails. Employer and employed are on that happy footing of equality and independence which should subsist everywhere. The company willingly pay to each what each can earn, and if at any time any workman has little to receive, he receives that little with the consciousness that he has himself to blame.

The machinery of the works. A detailed statement of the machinery of works so conspicuous and national is unnecessary. Suffice it to state that there are three powerful rolling-mills, in which, from old scrap iron, are manufactured the angle iron and plates which form the hulls of the ships. There are also seven large steam-hammers, under which the plates are hammered, as well as the heavy forgings for the keels, stems and stern-posts, and machinery of the gigantic ships. The machinery for planing, slotting, turning, boring, and otherwise fashioning these heavy forgings, is on the highest scale of efficiency and of the greatest power.

2.—VIEW OF THE MILLWALL COMPANY'S WORKS, MILLWALL.

Chapter VIII.

THE MILLWALL IRONWORKS AND SHIPBUILDING COMPANY (LIMITED).

Annual capability. These also are most important works. They possess a fine launching front of 1,500 feet on the River Thames, and in a time of pressure would no doubt turn out a very large amount of tonnage. At present the works are in course of development under the safe guidance of one of those self-made successful men of which England has so much reason to be proud. The Millwall Company have attained a high reputation for the production of rolled armour-plates. Leaving their achievements at Shoeburyness and Portsmouth unchronicled, they have quietly improved and perfected the "mixtures" for rolled armour-plates, until it may be said there is nothing left to be desired. Their Chalmers target is the most successful target that has yet been fired at, and from the new one, and the Reed's target, not less satisfactory results are anticipated.

The direction of the Company. Mr. George Harrison, the managing director of the company, probably possesses as great experience in the organisation of large bodies of workmen as any man in England. He has supervised and is identified with the completion of some of the most considerable public works at home and abroad. For years past it has been his occupation to set to work and keep at work several thousand workmen, so as always to be within contract time, and so that from every contract a profit may be realised. Higher qualifications for the discharge of the onerous duties of managing director of the Millwall Company could not be thought of. Under such guidance the economy of the Millwall Works cannot fail eventually to present a striking contrast to the economy of the dockyards as soon as the extensions and perfectings of the works now in progress have been finished. Already, of course, it does so. In the Millwall Works there are no starched officials; no Captain-Superintendent moving about at pleasure and interfering with impunity in matters of detail of which he knows nothing; no master shipwright charged with the duty of signing his name from ten till four o'clock at an annual salary of £600, with house, coal, and candle, besides a retiring allowance of £300, £400, or £600; and no leading men of gangs acting in the double capacity of sharers of the equal earnings of the gangs, and yet the check between the gangs and the employer. Mr. Harrison's assistant manager, Mr. Alexander, is an engineer of ability and experience. The naval architect of the works, Mr. Henwood, is well and favourably known to the profession. The gentleman in charge of the rolling-mills and forges, Mr. Hughes, has been long and most favourably known to the iron trade. Last of all, in the person of the secretary to the manager, Mr. Livingstone, is to be found an accomplished man of business. Such is the Millwall staff. Round the managing director the sort of men have been gathered that a man possessing great knowledge of the world would be supposed to choose. They are the best that capital can command and the skilled labour market of this great country could at the time supply. A company so directed cannot conceivably miss its mark—that, namely, of building ships of war and commerce, rolling armour-plates, and, if need be, forging cannon on more advantageous terms than they can be supplied by any public establishment of this or any country.

Employer and employed. There is another noteworthy point of difference between the labour system of the dockyards and that of the Millwall Company. The half-holiday has been established, the bell ringing at one o'clock on Saturday afternoon. For the convenience and comfort of those workmen who do not reside in the neighbourhood, a spacious dining-hall with a large stove has been provided. There is also a dining and reading room for the clerks, and before long there will be a library. Mr. Harrison is the patron of a workmen's rowing club, attends the matches, and bestows medals. A cricket club is encouraged in the same manner. So is a band. Such are the relations subsisting between employer and employed at the Millwall Works. The other day the rowing club had a match at Erith, and of the four or five thousand workmen employed at the works perhaps more than half the number were present. It is no degradation to labour at the Millwall Works. There the dignity of labour is recognised. Employer and employed stand on the broad level of buyer and sellers of labour. The Millwall Company have work to do, and the men are there to perform it. The one is necessary to the other, and the dependence is acknowledged. In the dockyards, on the contrary, labour is little better than a crime. The men cringe before their betters in the fashion of the black men of the Carolinas. When her Majesty last embarked at Woolwich for the Continent, the officials in their zeal penned the wretched dockyard workmen up until the embarkation was concluded, and that Royalty might not be seen by them, the windows were spread with whitewash. Worms are said to turn on those who trample on them, but in the face of such treatment it is asserted that it is within and not without the dockyards good workmanship is to be found, and therefore that ships of war, armour-plates, and even engines should be provided for the public service within the dockyards. They who believe the barefaced statement are only to be pitied. Human nature must necessarily recoil from the treatment received in the dockyards, and as necessarily give play to all its better instincts in works such as those of the Millwall Company.

Peculiarity of the works. The Millwall Iron and Shipbuilding Works are distinguished by the same peculiarity as the Thames Iron and Shipbuilding Works. Ships are not only built, but the iron entering into the construction of the ships is chiefly manufactured on the spot. The works are both ironworks and shipbuilding works. Nowhere else at home and abroad is the combination to be found. It is peculiar to the Millwall Company and to the Thames Company. They are establishments of infinitely greater national importance than all the seven dockyards, representing together a greater iron ship and armour-plate producing power than is at the present time possessed by the whole of France. To their support, not by subventions, but by employment in a straightforward business manner, public opinion cannot be too strongly called. Of two such establishments Englishmen may well be proud.

2.—VIEW OF THE MILLWALL COMPANY'S WORKS MILLWALL.

THE MILLWALL IRONWORKS.

The Model-room. Before making the round of the works with Mr. Livingstone, let us step into the model-room. This large and elaborately finished model was exhibited in the Exhibition, and represents her Majesty's frigate *Northumberland* in a sea-going state. It is the labour of some one hundred and fifty skilled workmen during three months, and is a marvel of beauty and completeness. Every part has been made to scale, guns, anchors, and wire rope, down to the tiniest blocks of the rigging, and the shot of the Armstrong guns. This is another model of some note, that of her Majesty's transport *Himalaya*. The *Himalaya* was built by the firm, and has long since proved one of the most serviceable ships on the Navy List. Her workmanship is perfect, and the rapidity and usefulness of her voyages are still cited in proof of the utility of transport ships belonging to the Crown, both in and out of Parliament. Round the room there are numerous models of lesser note of ships built and building.

The Building Yard. The building yard is separated from the forge and rolling-mills by a street, but the inconvenience is completely remedied by a tramroad and teams of stalwart horses. Entering the building yard gateway, the fine river frontage opens to the right and left. Here the *Great Eastern* was built. This slip to the left is that of her Majesty's frigate *Northumberland*, 400 feet in length between perpendiculars, and measuring 6,621 tons. At the bow of this noble ship the erections for the convenience of the workmen are commensurate with the magnitude of their task. These are the plate-bending rolls, capable of bending plates 16 feet in breadth; and these are the punching and shearing machines, and they are of great power. A steam-capstan hoists the beams into their places with a rapidity that is surprising. The *Northumberland* is building in an excavated slip, so as to allow the whole of the armour being fastened before the ship is launched. The next slip is occupied by a small screw trader of some 500 tons. The next to that has just had the blocks laid for a screw ship of 2,450 tons and 500-horse power, for the Royal Mail Steam-Packet Company. From the adjoining slip the fine screw ship *Baroda*, of 2,100 tons and 400-horse power, was launched the other day for the Peninsular and Oriental Company, and is now at the wharf of the Messrs. Humphrey opposite, receiving the engines and boilers. Passing by the landing wharves, we have next some half a dozen sailing ships of 1,500 tons each, and a screw collier of 900 tons. A little further the shipwrights are laying the foundations for a paddle-wheel ship of 1,700 tons and 400-horse power, for the Royal Mail Company, that being the second ship in hand for the same owners. Still beyond is the cupola corvette *Affondatore*, for the Italian navy, a ship of 3,000 tons, to be fitted with engines of 700-horse power, and estimated to steam seventeen miles an hour. This ship differs from those of our own navy in being armour-plated up to the main deck only, and in having the main deck covered with two-inch armour. The armament, within a fore and aft cupola on Captain Cowper P. Coles' plan, will consist of two heavy rifled guns; one gun in each cupola. The spars will be slight, sufficient merely to give steadiness to the ship at sea—a luxury that the Admiralty think proper to deny to the *Royal Sovereign* and the *Prince Albert*, the only two cupola ships yet in hand for our own navy. But the upper end of the shipbuilding yard has now been reached, and here are the extensive

2.—VIEW OF THE MILLWALL COMPANY'S WORKS. MILLWALL.

saw-mills and joiners' shops in a most accessible position for the workmen. Turning from these buildings, this further range of brick and mortar is the engine factory and foundries. It is needless to state that the machinery of these is adapted for all possible purposes, and that it is of the newest patterns and the greatest power. The engines at present in hand are a pair of 500-horse power, collectively, and a pair of 400-horse power, both for the Royal Mail steamers; a pair of 100-horse power for the screw collier, and a pair of 60-horse power for the screw trader. These factories in the course of a year could produce engines of an aggregate of 3,000-horse power. The foundries, two in number, are in full operation with castings for engines and the immense cast-iron columns for the railway bridges across the Thames. The minor departments we must pass over; the pattern, mould, and sail lofts, and the mast-making shed.

The Forge. Crossing the street from the building yard, the gateway opens on the extensive ironworks. The forge consists of six steam-hammers of the largest and most approved descriptions, furnished with steam and other cranes capable of turning out all descriptions of wrought-iron forgings up to 60 tons each piece. Here have been forged the stem and stern pieces for the *Northumberland*, the armour-plates for ditto, the large screw shafts and other parts of the engines for the *Northumberland* and other ships, the plates for Captain Inglis's two shields, as well as for the experimental targets of Mr. George Clark and Mr. James Chalmers, and the large stern frame for the Italian cupola ship building in the yard. In the smith's shops are four more steam-hammers of smaller dimensions, and above 100 smiths' fires. The process of manufacturing large forgings is as follows:—In the first place scrap iron, with an admixture of puddled bars to give the requisite quality, is piled upon a slab of wood to the weight of about 4cwt., and these piles are charged in a furnace heated to about 3,000 degrees Fahrenheit. When the pile has arrived at a full white welding heat, it is withdrawn and conveyed to the steam-hammer by a large tongs suspended from a girder in the roof, and is hammered into a thick slab of convenient dimensions. To convert these slabs into armour-plates or other forgings they are placed one by one on a large staff suspended from a powerful steam-crane, when the whole is welded by the steam-hammer into a solid piece; after successive laying on, heating, and hammering, which call for great judgment and skill on the part of managers and workmen, as well as faultless machinery, the glowing mass approaches its desired form and dimensions, and is finally removed to give place for another forging to be commenced. The cost of a steam-hammer of the first class, with cranes, furnaces, boilers, &c., averages £6,000. The quality of the iron in the armour-plates made here has been fully proved by the success of trial plates at Portsmouth, and more particularly in the two shields of Captain Inglis and Mr. Chalmers.

<small>Rolling-mills and Rolling.</small> There are two rolling-mills at present working at Millwall, one for angles and bar iron, and the other for plates and heavy bars, driven by a powerful pair of horizontal engines. In the former mill have been rolled the angle iron, 10 by $3\frac{1}{2}$— 8 by 8, &c., for the *Northumberland*, and flats up to 16 inches wide; in the latter mill, all the *Northumberland's* floor-plates, keel-plates, &c., and flat bars, 18 inches

wide and 5 inches thick. The process of rolling angle iron is very simple. Piles of bar iron are heated as described above and passed through grooved rollers, which reduce them to the required section, after which the ends are cut off by a steam circular saw, and the bar when cold is ready for the shipwright. The success of the operation depends upon the quality of the iron and the judgment used in turning the grooves in the rolls. The plant required is enormous, inasmuch as a separate pair of rolls is required for each section of iron to be made. The rolling of plates is very similar, only that the rolls have no grooves, being perfect cylinders. Many complaints have been made by ironmasters that the test imposed by the Government on ship plates is too severe, but, happily, such mixtures have been discovered as produce the required quality, and there is no fear of delay from this cause with the *Northumberland*. The "angle iron" and "plate mills" are furnished with several shearing-machines of great power and elegant construction, as well as a circular saw for cutting hot bars; all of these are worked by independent engines.

<small>The Armour-plate and Battery Mills.</small> The armour-plate and battery mills are now completed. The process of hammering armour-plates, as above detailed, is very generally considered to be the only method by which very heavy plates can be made, but there are grave objections to that system of manufacture, the principal of which are, the oft-repeated heating and constant hammering, which render the iron hard and brittle, and by no means suitable for resisting projectiles. None of these objections exist in the rolling of armour-plates, as this mode of manufacture requires only three heats at the utmost, and the pressure of the rolls, though greater

by far than the blow of a steam-hammer, is gradually applied; besides which the plate leaves the rolls at a good red heat, and, being allowed to cool slowly, will of necessity be rendered as tough as the iron constituting it will allow. The dimensions of this mill are so great as literally to make persons accustomed to the iron manufacture pause in astonishment. The fly-wheel, 36 feet in diameter, weighs upwards of 100 tons; the rolls are 8 feet long and 30 inches in diameter; and all the wheels, reversing gear, pinions, &c., are proportionately large and strong. The cost of this mill, with engines, boilers, furnaces, &c., will not be less than £100,000. There is still another rolling-mill in this establishment. This is called the "roughing mill," and is designed to supply the above three mills and the steam-hammers with moulds, tops and bottoms, rough bars, &c., from which are made the armour-plates, boiler-plates, angle iron, &c. The armour-plate mill and the roughing mill are driven by a powerful pair of horizontal engines, and are surmounted by a steam travelling crane for moving machinery, changing rolls, &c. A fifth mill is in contemplation, and plans are being made to erect it with little delay: it will be designed to make plates, bars, beams, solid rolled girders, &c., of the largest sizes, and its erection will widen the limits of the productive power of the company to an unprecedented degree. Attached to this department is a fitting shop, used more particularly for keeping the mills and hammers in an efficient state of repair.

Imperfection of this outline. Such is the imperfect outline of these great works, and the sagacious principles with which their economy is made to square. The great armour mill is capable of turning out 15,000 tons

THE MILLWALL IRONWORKS. 231

of armour-plates of any length or thickness in the course of a single year. The plate and angle-bar mills are capable of turning out 20,000 tons of plates and angle-bars annually, for ships, boilers, or bridges. The works cover 22 acres, and, as is well known, are directly opposite Deptford Dockyard, and accessible either by the river steamers of the Waterman Company, or by railway to the West India Dock station from Fenchurch-street.

Chapter IX.

THE MERSEY STEEL AND IRON WORKS, LIVERPOOL.

Position and extent of the Works.* These works have long occupied a prominent place in Liverpool. They were originated about the year 1810 by the late Mr. Ralph Clay, father of Mr. William Clay, the present managing partner of the company, and the late Mr. Roscoe. The original works were removed in 1862, the ground on which they stood being required by the Great Northern and Manchester, Sheffield, and Lincolnshire Railway Companies in forming their line of railway from Garston to the south end of the Liverpool Docks. The present works are close to the Harrington and the Toxteth Docks, and have direct communication with the railway just mentioned, a branch leading into the works, and communicating with the various seats of labour, so that every article produced can be readily transported by

---

* These and the remaining works of the same class are merely described, because it would have involved repetition to have treated them in the same manner as the Millwall Works and the Thames Works. Mr. William Clay, the managing director of the Mersey Ironworks, is eminently qualified to discharge the duties of a member of a Navy Council, should one be established on the plan of the India Council. Mr. Clay, Mr. Harrison, and Captain John Ford, as the members of a Navy Council, sanctioning all expenditure for *matériel*, and issuing all orders that involved expenditure for *matériel*, would soon rescue the navy from its present deplorable condition, and render it perfectly efficient for the purposes of war with a considerably reduced annual vote. Whether these gentlemen would undertake the task is, however, another matter.

1.—VIEW OF THE MERSEY IRONWORKS, LIVERPOOL.

railway, or water carriage, to all parts of the kingdom. The ground occupied is an irregular parallelogram, measuring 700 feet in length by 500 feet. It is divided into three unequal portions by two streets, one from west to east, the other from south to north; but direct and uninterrupted communication is maintained between all parts by means of tunnels. The northern half of the west portion is occupied by puddling furnaces, plate rolling-mills, forge furnaces, great steam-hammers, and powerful cranes worked by steam. It also contains a large, well-built, and suitable house 200 feet in length by 54 in width, and of proportionate height. In this there is a spacious suite of counting-house apartments and drawing-offices, together with a commodious dining-room, in which the clerks and drawing assistants dine regularly. Adjoining these offices, on the north, under the same roof, is an extensive engineering and fitting shop, supplied with apparatus of immense power, of the newest description. The southern portion of the west division is chiefly occupied by large rolling-mills and their furnaces. This division also contains a mill for the rolling of angle-iron of large dimensions. These mills are worked by a steam-engine of 250-horse power. The eastern portion of the premises is devoted to store-houses, refining furnaces and rolls, and a stock yard, in which the multifarious stores of a large establishment are classified and arranged, and where immense piles of iron—amounting occasionally to 8,000 or 9,000 tons of scrap, and an equally large stock of pig iron—are contained. It may be added that there are about 1,500 men and 50 horses regularly employed in the different departments, and eleven steam-engines of an aggregate of 2,000-horse power.

Puddle Steel and Furnaces. It would be foreign to our purpose to enter into any description of the process known in iron manufacture as "puddling," and we shall not refer to it further than to say that it is the operation by which cast or pig iron is brought into the condition of malleable or wrought iron, which is effected by two processes. The first of these is denominated the "refining" or second fusion of the iron, in the course of which a considerable amount of metallic and other impurities are discharged, and the refined iron cast into comparatively small cakes. Another process, which originated in Germany, has been greatly improved by the Mersey Steel and Iron Company. This is the peculiar metallic manufacture of what is called "puddle steel," which is found to possess all the properties of common steel, although produced at considerably less cost. From the improvements introduced into this manufacture, large quantities of excellent quality are produced. Besides the puddling furnaces, which are arranged in avenues resembling streets, there are also in their immediate vicinity several furnaces specially adapted to the heating, welding, and working up of scrap iron; and in all, the processes of manufacture are carried on day and night, ceasing only on Saturday afternoon, to be resumed on Monday morning. By this continuity of operation, the adoption of the most improved methods of working, and the employment of the best and most efficient machinery, the Mersey Steel and Iron Works are capable of turning out 600 tons of malleable iron and puddle steel weekly.

Furnaces and Machinery. Immediately to the north of the puddling furnaces are numerous large furnaces adapted to the heating of the metal preparatory to its being passed through the plate-rolling mills, which in this division of the works are driven by a steam-engine of 250-horse power. The motion of the fly-wheel connected with this engine is somewhat startling. The wheel is 35 feet in diameter, and weighs 60 tons; and while at work it makes thirty-eight revolutions in the minute. The rolling-mills are suited to the production of plates up to 2 inches thick by 5 feet 6 inches wide, and the rolls to the drawing of rod or bar iron. Connected with the same motive machinery, and contiguous to the rolling-mills, is a trimming instrument, or cutting shears, sufficiently large and powerful for paring and trimming iron plates of large dimensions; besides other apparatus.

The Forge Department. The space to the north immediately adjoining these heating furnaces and rolling-mills, containing an area of not less than 42,140 square feet, but completely roofed in, is appropriated to the operations of the forge department. This large working space is completely filled, but not crowded, with steam-hammers, some of which are of great power, welding furnaces of extraordinary capacity, and powerful steam-cranes, by which immense forgings are lifted and swayed about with an alacrity and precision that might well be considered wonderful, even with comparatively small work, but which, applied to the enormous masses brought under the forge steam-hammers, is really marvellous. What adds to the astonishment of the spectator is the regularity and self-possession with which the forgemen pursue their labour. To the International Exhibition the company contributed between 150 and 200 tons of forgings, including some of the largest cast-iron fabrics ever made. Among these were two crank-shafts for marine engines. One was exhibited in the finished state by Messrs. Penn and Son of London, for whom it was forged; the other by the Mersey Steel and Iron Company, in the rough—that is, just as it left the hammer. This last was 30 feet in length by 20 inches in diameter, and had two crank-blanks. Each of these blanks measured 5 feet on the shaft, and was 4 feet 4 inches in depth. The distance from centre to centre of the crank-blanks was 10 feet; the blanks to be so cut out as to give a 4 feet 4 inch stroke. The weight of the forging was fully 25 tons, and the whole was completed in between five and six weeks. These immense shafts were forged respectively for H.M.S. *Northumberland* or *Minotaur*, and the steam-ram *Achilles*, the engines of each of which are of 1,350 and 1,250 horse power respectively. Another great forging of the firm is the stern-post of the armour-plated steam war-ship *Agincourt*, now building by Messrs. Laird Brothers, Birkenhead. This is 42 feet in length, and, with its sole and shaft-boss, weighs 40 tons.

<small>Wrought-iron Cannon manufacture.</small> So long ago as the year 1845 the attention of the Mersey Steel and Iron Company had been directed to the construction of wrought-iron guns, forged in such a manner as to be perfectly homogeneous in texture all through the structure, and in that year they forged, in every respect successfully, a large gun for the United States steam frigate *Princeton*. This gun was 13 feet in length from breech to muzzle; and previous to being turned and bored, it weighed 11 tons 3cwt. 2qr. 11lb. The length of the bore was 12 feet by a diameter of 12 inches. The boring and turning revealed neither flaw nor fault in the texture of the gun, and when finished it weighed

7 tons 17cwt. 1qr.; proving capable of throwing a spherical shot of 219lb. This gun is still in existence, and in perfectly good order. In 1856, in answer to the allegation that malleable iron ordnance of large calibre could not possibly be made, the company forged, finished, and presented to the nation their world-renowned "Monster Horsfall Gun." This, as is well known, is of the following dimensions:—Length from breech to muzzle, 16 feet; length of bore, 13 feet, by a diameter of 13 inches. The finished weight of this enormous piece of ordnance is 24 tons 3cwt. 2qr. 21lb.; when the forging was completed, and before the gun was turned or bored, it weighed 28 tons 1cwt. 3qr. 21lb. In this, as in the previous instance, the processes of boring and turning showed that the forging was perfect and the metal thoroughly sound and tenacious; further proof of this being furnished by the repeated tests to which the gun was put by the Board of Ordnance, all of which it passed with the most unimpeachable success. Again, in 1861, the company forged another immense gun, which was exhibited at the Exhibition in the following year, under the name of the "Prince Alfred Gun." This gun is 12 feet 6 inches in length from breech to muzzle, has a bore of 10 feet 6 inches long by 10 inches in diameter, and in its finished state weighs 10 tons 15cwt. 2qr. 14lb., but as it was taken from the forge, before it was turned or bored, it weighed 13 tons 10cwt. Like the others, the "Prince Alfred Gun" also passed successfully through all the initiatory and other ordeals which ordnance of the highest quality and character are presumed to undergo. In addition to the guns named, large numbers of wrought-iron field and siege guns of various dimensions have been most successfully manufactured in the works.

The Forge appliances. Let us now turn to the mechanical appliances of the forge. The furnace is 17 feet, or, including the furnace-neck, 24 feet in length by 8 feet wide, and is furnished with a draught flue 56 feet in height. The crane is 24 feet in height above the ground, but as it is sunk 12 feet in the solid rock, the entire height of the upright shaft is 36 feet. This crane is a double rectangular tubular one, constructed entirely of plate iron rolled in the works—each entire side of the crane-formed tube being of one plate, a marvel both for form and magnitude. On the upper side of the horizontal arm there are two travelling carriages for the purpose of giving a backward or forward motion to the mass of metal held in suspension. The crane has also a rotatory motion of the whole fabric, for swinging its head round between the hammer and the forge, and a winding-up motion for lifting the mass to be submitted to the action of the hammer. By the combination of these three perfectly distinct and independent motions, which, however, can all be applied simultaneously, the largest masses can be lifted, poised, or laid down at any point with the nicest accuracy. The whole weight of the crane is upwards of 50 tons, and it is capable of lifting a mass weighing 120 tons.

Stupendous Hammer. The next object of interest is a stupendous hammer, by which masses of iron are fashioned as easily as in most cases a piece weighing not more than pounds for tons is manipulated. A glance will show that it is a complicated as well as a powerful implement. The width between its upright supports is 14 feet 6 inches; the weight of the piston and hammer is fully 8 tons, and when employed working it has a fall of nearly 7 feet. The piston and projection rod, which are in reality one, is 15 inches in diameter by 7 feet 6 inches in length, made and forged at the works. The anvil-block of the hammer, which is deeply bedded in the solid rock, is 9 feet square on the base, and weighs 32 tons 15cwt. The height of the hammer and its frame is 23 feet, and the absolute weight of metal in the apparatus, including bed-plates, framing, and anvil-block, is fully 70 tons. As may readily be supposed, the full power of such an implement is very great. The monster hammer of the works is of the following dimensions:—The width between the upright supporting columns is 25 feet. The weight of piston and hammer, a solid mass of iron, is 15 tons; and the piston and projection rod is 20 inches in diameter by 15 feet 2 inches in length. When in operation this enormous striking mass has a fall of about 9 feet, imparting a blow of almost incalculable force, which causes the solid rock in which the machine is securely fixed to vibrate and tremble as if shaken by an earthquake. The anvil-block of this ponderous and powerful implement, is $10\frac{1}{2}$ feet square on the base, $6\frac{1}{2}$ feet in height, and weighs 62 tons,—probably one of the largest masses of cast iron in the world. The total weight of this huge instrument of power, including bed-plates, framing, and anvil-block, is 130 tons; and so admirably adjusted are all the parts of the implement and its apparatus, that, while it can be made to give a blow of inconceivably destructive force, it can also be made to strike as gently as the tapping of a lady's fan. Like the one previously described, this immense hammer is also supplied with a heating furnace of proportionate size and heating power, with two cranes of like character, so arranged with relation to each other, and to the large furnace and hammer, that their action and power can be combined. Besides these two steam-hammers, the forge department contains other seven. Of these, two have hammers weighing 6 tons each; one weighing 5 tons; one 3 tons 10 cwt.; one 3 tons; one 2 tons 10 cwt.; one 1 ton 10 cwt.; and in the smithy attached to the forge department there is yet another of still smaller dimensions.

<small>The Engineering Shop.</small> Immediately adjoining the forge department—indeed, it may be said, forming a complementary part of the forges—is an extensive and fully equipped fitting or engineering shop. This forms a parallelogram, the area of which is 200 feet by 54 feet, with great height of walls. The whole of the lower or ground floor of this spacious apartment is devoted to the trimming, planing, boring, or turning of the gigantic forgings which are produced in the establishment, so that all the work contracted for by the company may be satisfactorily completed. As an instrument of application in the way of lifting and carrying, common to all the work brought into this branch of the establishment, is a high-level railway, supported on lofty pillars, running from end to end of the shop. On this railway a self-acting travelling crane is placed, of large size, and capable of carrying 30 tons weight. This crane is worked by a band, and is so easy of management that a boy can move it with its load with perfect ease and in any direction. The central portion of the floor of this large and commodious workshop is occupied by a gigantic planing-machine, which is nearly 12 feet wide, and is 40 feet long in the travelling table. Besides being capable of planing a surface of this size, the table is furnished with an ingeniously-contrived means of regulating the apparatus so that

1.—VIEW OF THE MERSEY IRONWORKS, LIVERPOOL.

forgings of any species or degree of curvature may be planed and trimmed with the nicest accuracy. In this way the stern, 18 tons in weight, of the armour-plated war-ship *Agincourt* was planed after it had been bent; and the different grooves and rebates necessary for receiving the armour-plates and other necessary portions of the ponderous fabric were cut out and adjusted with the most perfect success and accuracy. The south-eastern portion of the shop is devoted to the working accommodation of two immense turning-lathes. One of these is 6 feet 3 inches high in its centre point, thus giving the means of turning a mass 12 feet 6 inches in diameter; the other 5 feet 3 inches high in its centre point. These two gigantic lathes are furnished with moveable rests, which can be shifted so far from the head, or fastening disc, as to take in a shaft 65 feet in length; and it is no unusual thing to see one or other of these lathes giving rotatory motion to a 25-ton crank-shaft, or one of the ponderous guns for the manufacture of which the Mersey Steel and Iron Works have gained a world-wide and well-deserved reputation. The south-western, or opposite portion of the shop, is occupied by lathes of smaller dimensions and less power than the two just mentioned, but still of such magnitude as to entitle them, even there, to the appellation of very large, while in most other establishments they would merit the title of gigantic. Immediately to the north of the vast planing-machine is one of recent contrivance, adapted to the trimming and cutting of armour-plates. This powerful implement of labour is capable of taking in armour-plates 20 feet in length by 4 feet in width, of any thickness, cutting and grooving the edges, which it trims and fits with accuracy.

The Rolling and Armour-plate Mills. About one-third of this large shop, at its north end, is covered by a floor at about 20 feet above the ground. This upper floor is liberally supplied with the lathes, drills, slotting-machines, and other implements of an extensive and well-furnished engineering establishment. The next division of these important works is the south-western portion. It is 200 feet in length from north to south, by 295 feet in width from west to east, divided exteriorly from the other part of the works by Egerton-street, but immediately connected with the works by a tunnel under the street. This section is chiefly taken up with rolling-mills of different capacities, suited to the production of widely different classes of work. Conspicuous among these is the powerful mill for rolling armour-plates intended for ironclad war-ships, and also for casing some of the more important of our coast defence fortresses. The productive power of this mill is astonishing: it will manufacture armour-plates from 20 to 40 feet long, 7 feet 6 inches wide, and of any thickness that may be wanted. Of these enormous plates it can produce four each day, or twenty-four in a week. The mill consists of numerous huge but well-arranged portions, the motions and magnitude of which convey a very lasting impression to the mind of the spectator. The apparatus is supported on a framework composed of massive ironwork bedded deeply in the solid rock, and securely fastened to a well-constructed foundation framing of oak beams, which measure 24 inches by 22 inches on the sides. These are of the best Quebec oak, about 40 feet long, in two lengths, the joints well scarfed and bolted; the ends lever-jointed and granted into the carefully excavated rock; and the breadth of this wooden frame is 11 feet. On these immense

THE MERSEY IRONWORKS.

oaken beams the cast-iron bed-plate is firmly bolted. It is 36 feet in length by 9 feet in width, and 6 inches thick, weighing 35 tons. Very heavy and thick uprights are securely fastened to this bed-plate, at conveniently arranged distances for supporting the different portions of the moving machinery. The fly-wheel is 25 feet in diameter, 15 inches broad on the rim-face, and weighs between 50 and 60 tons. Working on the same shaft with the fly-wheel is the driving-wheel, 3 feet 6 inches in diameter, revolving, of course, at the same rate as the fly-wheel. This works into two cog-wheels, each 8 feet in diameter, 18 inches broad on the face, and weighing 10 tons. These again work on their respective shafts another such cog-wheel, 12 feet in diameter, 18 inches broad on the face, and weighing 12 tons; the latter being supplied with reversing crabs, and working in sills weighing 7 tons each. The shafts to which these are affixed are 20 feet in length, and between them they weigh 13 tons. Attached to these shafts, by means of pinions and spindles, which weigh 5 tons and work in housings of 7 tons weight each, are the rolls. There are two immense cylinders, 8 feet long in the barrel, with a diameter of 2 feet 6 inches; the pair weighing not less than 22 tons. They work in housings 11 feet high, 7 feet 6 inches broad, and weighing 11 tons each. The preceding constitute a few of the leading details connected with this stupendous working tool, which contains in all a weight of metal exceeding 300 tons! Yet all so perfectly arranged and well contrived as to work with the greatest steadiness and the most perfect regularity.

The Heating Furnace. Immediately behind this massive congeries of machinery is built the heating furnace. This is a low building of large area connected with two flues, with a length of between 50 and 60 feet to secure a strong and perfect draught. The furnace front is 20 feet long, and its interior area measures 14 feet by 9 feet on the floor. From the furnace door a railway is laid on a slight inclination towards the rolls, and on this railway a carriage of the strongest and most powerful construction travels upon low wheels, carrying the enormous masses of iron, heated to the most dazzling white heat, down to the rolls through which its fiery burden is to be passed, and by which it is to be fashioned into the ponderous casing-plates which constitute the protection of our ironclads against the crashing power of hostile artillery. On the opposite side of the rolls, and exactly *vis-à-vis* to the carriage spoken of, is another of similar dimensions and formation to receive the extruded mass and pass it back again, so that it may be submitted to the requisite amount of pressure which shall ensure to it perfect solidity and thorough homogeneousness of texture. When this has been accomplished, the large plate is removed to a convenient spot for cooling, after which it is conveyed into the store pile, while the implements of its manufacture are again put through their respective rounds of action, to produce a similar tangible and ponderous result. After these massive plates have been thoroughly cooled, they are conveyed to the fitting-shop, where, by means of the cutting and trimming machine to which allusion has been already made, they are trimmed and fitted to the exact form and dimensions they are intended to assume. Besides the ponderous rolling-mill, the division of the Mersey Steel and Iron Works now under consideration is fitted with a vast rolling-mill for the purpose of rolling bar, rod, and angle iron of large dimensions; and to the south-east of the armour-plate rolls there is a set of rolls for drawing smaller work. As to the general efficiency and excellence of the armour-plates produced at this establishment, it may be sufficient to state that several of the sample-plates have been subjected to the severest tests, and have come through the ordeal in such a manner as to elicit high commendation from the judges appointed by the Lords of the Admiralty. It may also be worth while to state that at the Great International Exhibition of 1862 they exhibited among their other forgings one of the largest marine battery plates ever forged, and much beyond the size of those ordinarily forged. They also exhibited the first battery iron-plate target which ever was fired at, and which was broken by a shot from the stupendous wrought-iron gun which they made and presented to Government; and along with the target they also exhibited the shot that broke it.

Employer and employed. Having thus cursorily glanced at the machinery and what may be called dead plant of this establishment, with a view to estimate its great power and admirable adaptation to the production of the largest and highest class work connected with iron manufacture, it might appear that our descriptive task was accomplished, but this could scarcely be done without pointing out the important fact that the company is in all respects on the best terms with its large and intelligent staff of *employés* and numerous workmen. Among these, embracing, as it necessarily does, men of all ranks and of very varying degrees of intellect and talent, the benevolent management of the company has been conspicuously shown in its establishment of, and the fostering care it has bestowed on a sick and burial society connected with the works, managed entirely by the workmen. This society holds periodical meetings, the annual one being generally presided over by the managing partner of the firm, and concluded by one of those genial festivities in which all classes of Englishmen delight to mingle. The company have also established and maintain a well-appointed and efficient reading-room for the use of the workmen.

Chapter X.

THE ATLAS WORKS, SHEFFIELD.

Origin and progress of the Works. Less than ten years ago Mr. Brown was the occupier of premises in Furnival-street, of very moderate dimensions, noticeable only for the activity of their operations and the fact that they were the works of a young man who had risen from the humblest position, and owed his success entirely to his own energy and enterprise. On the failure of Messrs. Armitage, Frankish, and Barker, Mr. Brown became the purchaser of their newly erected works, adjoining the Midland Railway in Saville-street East. The old name, "Atlas Works," was transferred to the new premises. The site comprised nearly $3\frac{1}{2}$ acres—the present area of the works south of the railway—and though probably less than half that area was covered with buildings, the removal to premises of such magnitude was deemed by many as a bold step. The works, however, speedily proved too limited for the expanding trade of the new proprietor. On New Year's Day, 1858, Mr. Brown started a newly erected rolling-mill, 63 feet square, on the west side of the works, giving a dinner to 400 of his workmen. This, though not the first extension, was the most notable that had yet been made. It was to increase the facilities for carrying on the branches of manufactures already in operation, not to introduce new ones. At this period Sheffield manufacturers, even of springs and other railway material, received their supplies of iron from remote places in a partially manufactured state. None of them purchased it in the pig and manufactured it for themselves; and even such of the processes as were performed in Sheffield were done so at different places. The iron was tilted at one place, forged at another, and so on, reaching the manufacturers in the state ready for the ultimate processes for which it was required, after going through half a dozen distinct and sometimes distant works of different manufacturers. The disadvantage of this state of things was twofold: Sheffield manufacturers were paying men at a distance for smelting from the pig an article of which they were using immense quantities; and after the iron did reach Sheffield, it was transported at great expense from one place to another to undergo the necessary preliminary processes. This state of things was not longer to continue. Mr. Brown had no sooner got his large rolling-mill into operation than he embarked in an enterprise which resulted in the introduction of a new and extensive branch of manufacture in Sheffield, and the concentration within single manufactories of the various processes of iron manufacture. In April, 1858, Mr. Brown completed the erection of six puddling furnaces, with ball and other furnaces, Nasmyth hammers, and all the appliances for the manufacture of iron from the pig, and its conversion into the different qualities of steel for the multifarious purposes for which it was required. Mr. Brown had now also filled with extensive railway spring and buffer shops, smiths' shops, casting shops, rolling-mills, furnaces, forges, &c., the whole of the $3\frac{1}{2}$ acres of land south of the Midland Railway. His works were among the largest in the town, and extensive enough for the enterprise of any ordinary man.

He, however, was surrounded by competitors of rare tact and capacity, and of enterprise equal to his own, and emulating their great successes, his ardent nature disdained the limits of even such ample proportions as the Atlas Works could then boast. Immediately after the French successes in armour-plated ships were forced on the serious attention of the Government and people of this country, Mr. Brown entered with his accustomed earnestness into the new branch of manufacture which the science of war had created. Having taken upwards of 10 acres of land between Carlisle-street East and the Midland Railway, immediately opposite his existing works, Mr. Brown traced out, on Whit-Monday, 1859, the foundations of an extensive mill for the manufacture of armour-plates. Forged armour-plates were first tried, but the experiments undertaken by Government demonstrated the necessity of some process by which greater tenacity and power of resistance could be obtained; and about the 20th of March, 1861, Messrs. Brown and Co. (for Messrs. Bragge and Ellis had meanwhile entered the concern as partners) commenced with encouraging success the rolling of armour-plates. Great obstacles had to be encountered in a branch of manufacture so entirely new, but Messrs. Brown and Co. have succeeded in overcoming all.

The Old Planing Shop, &c. The "Old Planing Shop" is a room 56 by 96 feet in extent, abutting north on Carlisle-street East, and west on the works of Mr. Bessemer. When the manufacture of armour-plates was undertaken at the Atlas Works, this capacious shop, now regarded as a very small and insignificant part of the premises, was fitted for the planing and "slotting" of armour-plates. In consequence of the erection of a new and much larger planing room near the eastern end of the new works, that shop is now partly filled with machinery for rolling railway rails, and kindred purposes. Flanking the shop on the south are ten large converting furnaces, each capable of making thirty-five tons of steel at a time. Beyond these are mills fitted with the most approved machinery for rolling steel plates and steel for railway and other springs. Adjoining the mills is a tilt for hammering steel for tools, chisels, and other purposes for which rolled steel is not fitted. Huge hammer heads and faces are fastened on the ends of rough beams of wood fixed in a horizontal position. The beams are worked by machinery with such rapidity that when in full operation the lesser of the tilts delivers 300 strokes per minute, the larger ones striking from 100 to 150 times per minute. A spring fitting shop, 80 feet by 62 feet, occupies the space between the tilt and the railway, completing the division of the works between the broad tramway running straight across from the Carlisle-street entrance to the railway. Crossing this tramway, the visitor reaches a lofty mill fitted with two immense caldrons for making cast steel, on Mr. Bessemer's interesting process. Steel thus cast is used for axles, railway rails and tires, piston-rods, guns, and a great variety of other purposes. Beside the mill are the beautiful engines that supply the draught of air blown through the molten steel in the caldrons—an operation peculiar to the Bessemer process. Immediately the visitor passes round the front of this engine-house, to the right there is the "Old Armour Plate Mill," 360 feet long, roofed by two spans each 75 feet wide. This mill, which, except a portion of the south side, recently built as a puddling forge, for making cast into malleable iron, was the first part of the new works erected, occupies the centre of the premises north of the railway, extending from the railway about two-thirds of the distance to Carlisle-street East. As the eye of the visitor wanders curiously over its ample dimensions, there is seen, in apparent confusion, but really in the most admirable order, sets of armour-plate rollers, domineered over by tall, strange-looking curved cranes, which ever and anon snatch the seething, newly-rolled armour-plates from the mill floor and swing them upon trucks for removal to the planing-room; Naylor and other steam-hammers, six tons in weight, and capable of striking with a force of sixty tons; and tilts with leviathan heads and enormous fly-wheels revolving at almost lightning speed. Surrounding each set of rolls, each hammer and each tilt, are capacious furnaces, glaring with eyeballs of fire upon the visitor, as they heat the armour-plates or the cast-steel guns, shafts, and piston-rods, for rolling or forging, as the case may be; or as, in the new portion of the mill, they smelt the pig iron into a glowing fuming liquid, into which the puddler from time to time thrusts his huge tongs, and having by dint of turning them round and round succeeded in rolling together, like a snowball, a rough mass of molten metal, hurries with it to the hammer, where it is beaten into blooms (squares) or slabs, throwing out in this process of "shingling," as it is called, showers of red fragments. Interspersed with rollers and hammers, tilts and furnaces, are machines for cutting into shape the slabs of which armour-plates are constructed, and a variety of other appliances required in carrying out the diversified processes which are continually going on night and day. The numerous chimneys and brick domes surmounting the furnaces form a feature in the aspect of the mill, which, amid the clang of wheels, the dull reverberations of hammers, the hissing of furnaces, the screeching of steam pipes, and the clatter of tongs and other implements in the hands of the groups of swarthy workmen scattered over the place, present a scene of activity and of the operation of the vast forces of nature evoked, and directed by man's intelligence in aid of the purposes and necessities of civilisation.

The New Rolling-mill. Passing out of the old rolling-mill at the east end we reach the new rolling-mill, which is 250 feet in length and 150 feet in width. In the centre of the room is the engine-house, containing two engines of the nominal power together of 300 horses, but really of far greater power. Projecting from the engine-house are the enormous fly, cog, and other wheels by which the machinery is driven. Two sets of rolls, 32 inches in diameter and 8 feet long, are erected across the centre of the mill, from the driving machinery to the north side. The two rollers of each set are, of course, fixed one over the other, the armour-plates being crushed between them in the process of rolling. Huge as are the rolls, they have to be so fixed that, while giving a power of compression limited only by the strength of the iron framework in which they are set, they can be easily and instantly drawn close together, as the plate becomes thinner and thinner with each rolling. This object is accomplished by means of ponderous balances, raised or lowered at will by the turning of a large screw at each end of the wheel. There are two enormous furnaces on one side of the rolls, and one on the other, at the distance of about 60 feet, and it is in these furnaces that the iron plates are heated preparatory to being rolled. The floor of the mill is flagged with square iron plates, and declines considerably from the furnaces to the rolls. The plates are conveyed from the furnaces to the rolls on long iron trucks or lurries. The wheels of the lurrie run in grooves from the furnaces to the rolls, for the purpose of securing accuracy in depositing the plates in the rolls. The new mill is at present far less thickly studded with machinery than the old mill. Indeed, the part of it which is to be used as a rail mill is not yet fitted up. It is, however, a lofty, noble-looking room; and, profiting by their experience in the old mill, Messrs. Brown and Co. have so constructed it as greatly to facilitate operations.

<small>The Planing and Slotting Shop, &c.</small> Beyond the new mill is the new planing and slotting shop, 220 feet by 75 feet in size, containing eight planing-machines, and five single and one double slotting-machines. These machines are of the most improved construction, and by the simplest arrangement can be made not only to plane the plates into squares with perfectly even surfaces, but also to plane the edges into as equally perfect a semicircle, and also plane one end of the plate thinner than the other, graduating the thinning with perfect evenness and accuracy from end to end—these latter processes being necessary to fit the plates to the curving and sloping sides of the ships. Preparations are being made for occupying the space of ground from the new planing-shop to Hallcarr-lane, which forms the eastern boundary of the works, with a new hammer shop, and a converting foundry for casting iron generally. A broad road runs along the side of the mills and shops from the old planing-shop at the western to the new planing-shop at the eastern extremity. The space between this broad road and Carlisle-street East is occupied with buildings

in course of being fitted for refining iron, and with joiners' shops, yard for storing pig iron, fitting, repairing, and casting shops—all repairs of machinery, &c., being done on the premises, where also some of the machinery is made. A tramway with iron rails has been laid down on the road close to the mills, as also across the works at several points, for facility in moving armour-plates and material generally from one department to another, and from the planing-shops to the railway. A moveable steam-crane plies on these tramways, facilitating to a marvellous extent the loading and unloading, and other operations. Vast as is the extent of the works, every department is fitted up with machinery to almost a lavish extent. Much of the machinery is of the most massive character; the more valuable parts having been obtained from the very best makers in the country, including Sir William Fairbairn, and other celebrated engineers and manufacturers. From end to end of the premises, the works are firmly and substantially built, and the number of men employed is not less than 2,500. There are also 45 engines, some of them 150-horse power each, on the premises, 60 puddling furnaces of the largest capacity, and a still greater number of converting, heating, and other furnaces. The weight of material turned out per week cannot readily be ascertained with any approach to accuracy. Some idea of its enormous extent may, however, be formed from the fact that the weekly consumption of coal exceeds 2,000 tons, and that the outgoings are over £1,000 a day.

What the Admiralty overlook. What the Admiralty overlook in proposing to establish such works as those of Messrs. John Brown and Co., is the fact that the

Atlas Ironworks are not the creation of a day, but of a life. True, the great extensions of the Atlas Ironworks are of yesterday, but Mr. Brown nevertheless perceived their utility, and the mode of judicious management, before the ground was purchased, much less built on. He and his partners, by a long course of business training, knew exactly what would answer. The Admiralty, on the other hand, know and see nothing but the great results of the Atlas and other works, and those results they envy. They see what Messrs. John Brown and Co. can do, and they determine on doing likewise. They count as nothing the experience, tact, liberality, power of combination, supervision, direction, and we do not know what besides, which together form the great conditions of Messrs. John Brown and Co.'s success. When the firm entertained their noble visitors the other day, who were always present to Mr. Brown's mind? Not the visitors, nor his partners in business, but "his brave workmen." Mr. Brown was constant in his admiration of the "brave workmen," who had heated 20 tons of iron, dragged it from the furnace, and passed it in an instant through the rolls into an armour-plate 40 feet in length, 4 feet in width, and $4\frac{1}{2}$ inches in thickness. The like had not been done before, not in England only, but in the world. Twenty tons of iron in a white heat were for the first time seen, faced, and handled as quickly as seen, and what has been once done can be done again. With his "brave workmen" Mr. Brown can roll plates such as no ships afloat can carry comfortably, and such as no artillery yet known to science can penetrate.

Chapter XI.

THE PARKGATE IRONWORKS, YORKSHIRE.

Origin and progress of the works. The Parkgate Ironworks are situated near Rotherham, in the West Riding of Yorkshire, on the borders of Derbyshire. In 1839 these works were erected for the manufacture of bar and sheet iron, and the first considerable extension dates from 1845, the great year of railways. During that year a rolling-mill was added to the works, capable of producing 2,000 tons of rails monthly; and it may be remarked that the rails turned out by this mill are hammered before rolling. At a later period, when iron ships and iron bridges became popular, the proprietors, Messrs. Samuel Beale and Co., were induced, by the demand for large and heavy plates, to lay down a plate mill of the largest description that had then been erected, and at this mill all the plates of the *Great Eastern* were rolled, including those of the stern. Again, in 1856, during the war with Russia, when the Government ordered several floating batteries to be built and covered with $4\frac{1}{2}$-inch plates, Messrs. Beale and Co. were applied to by Messrs. Palmer Brothers, the eminent shipbuilders on the Tyne, to know whether plates $4\frac{1}{2}$ inches thick, and weighing about 3 tons each, could be rolled, as up to that time there had been no such heavy plates rolled. Messrs. Beale undertook the work, and after many experiments and great outlay succeeded in producing armour-plates up to six tons weight each plate. For nearly five years Messrs. Beale and Co., being the only manufacturers of rolled armour-plates, had to contend single-handed in support of rolled, against hammered plates, and at the comparative trials at Woolwich and Portsmouth it has been proved to demonstration that on the whole rolled armour-plates are to be preferred. Contemporaneously with the expansion of the Parkgate Rolling-mills, large planing and slotting machines for the purpose of finishing the armour-plates had to be procured; and recently, when the Admiralty considered it advisable to have the armour-plates bent while hot, machines for the purpose of bending them have been erected, and plates can now be bent as they come from the rolls to any desired curve. Plates so bent have, as a matter of course, stood the tests much better than plates bent cold.

Extent and facilities. The works cover a large area, and possess the facilities of canal and railway, and adjacent mines. Coal, the consumption of which is something enormous in all ironworks, may be said to exist within the Parkgate Works, and therefore has the advantage of being near the mills and furnaces. This is of great importance, although its value cannot very well be estimated. For instance, in the case of comparatively remote works, there is a great relative outlay under the head of the carriage of the finished products to market, whether the latter outlay is supposed to come out of the pocket of the seller or the buyer. But inasmuch as the bulk of the finished iron product is less than the bulk of the coal necessary to produce it, it follows that, other things being equal, it is more profitable to establish ironworks beside the coal than beside the market in which the finished iron product is required. The advantage, however, may or may not be enjoyed by the buyer just as the seller is or is not required to yield it. Passing by this consideration, let us now direct attention to the completeness of the Parkgate Works, as well as to the care with which the productiveness of labour is kept in view throughout. We enter the works by the coal tramway. On the right is the foundry, and on the left the fitting shop. The railway passes the foundry, enabling it to supply pig iron and coal. The wants of the foundry attended to, the railway, which up to that point is a single line, branches off in four directions for the supply of the blast and puddling furnaces. This arrangement is admirable, as from this point the coal is taken direct to the various furnaces, which are thus kept blazing to their full capacity, no time wasted, and no superfluous unassisted labour being employed. Immediately in the rear of the second line of puddling furnaces, a steam-hammer in the centre of an open space is flanked by engine-houses with engines driving slabbing rolls, puddle-bar rolls, tilt-hammer, &c. Outside this machinery, and extending backwards to the coal railway, all the necessary conveniences of a great establishment are grouped,—smiths' shops, warehouses, pattern lofts, fitting shops, sawpits, carpenters' shops, &c.,—that the sometimes trying race of commercial competition may be run by the Parkgate Ironworks with the least possible dead weight. Such is the imperfect outline of one of the important branches of the Parkgate Works. Crossing a turnpike-road, this extensive range of buildings is the rolling-mill establishment,

THE PARKGATE IRONWORKS.

and round it is a line of railway with double branches to two powerful cranes, and a branch to the pig iron yard, while the main line extends to the armour-plate shop, the latter a new outlying extension of the works to the left. Facing the turnpike there is a line of furnaces; and, forming two more sides of a square, additional lines of furnaces extend to the right and left. In the hollow of these three sides of a square is the engine-house, from which No. 1 and No. 2 rolling-mills are driven. Shears, cranes, and the other conveniences of an armour-plate mill abound, and the armour-plate capacity of both mills must be very great. Altogether this branch of the establishment is as complete as the first. All that capital, enterprise, and intelligence could do has been accomplished, so that the Parkgate Ironworks may produce as cheaply, if not cheaper than their rivals. There is a third department of the Parkgate Works— the rail mill department, round which a line of railway runs with convenient sidings. But of this it is unnecessary to speak. It is a complete establishment in its way, and has long been in successful operation. Altogether, connected with the Parkgate Works are four blast furnaces—one at Parkgate, two at the Holmes, and one in Derbyshire—all supplying good and suitable iron for the requirements of the Works. There are about 60 puddling furnaces, and 30 mill furnaces; three 6-ton tilt hammers, and one 5-ton steam-hammer; and in addition to the rail mill and boiler-plate mill there is a merchant mill for making bar and hoop iron.

The lesson taught by the Works. The lesson, in a word, taught by the Parkgate Works, is the same as that taught by the others. Private enterprise is equal to the supply of the wants of the public service in any form, and on better terms than the public establishments. Between the Messrs. Beale and their workmen a feeling exists that experience proves to be impossible in the dockyards. In the service of considerate masters workmen exert themselves to an extent that is surprising, and become as watchful of the interests of their employers as the latter can be. And the necessary consequence of such a relationship subsisting between employer and employed is the payment of a rate of wages that practically is unattainable in the dockyards. What men in the private establishments earn, their employers can well afford to pay, while in the dockyards it would perhaps be an impropriety were the wages of an humble but skilled hammerman the £900 per annum that is said to be paid to one of the hammermen in the service of the Mersey Steel and Iron Works. In the private establishments the phrase "limited earnings" is unknown, and from year to year there is no fluctuation from task and job to day wages and back again in a circle. The rule of the private establishments is to encourage workmen to earn all they can.

9.—VIEW OF MESSRS. MAUDSLAY, SON, AND FIELD'S WORKS, LAMBETH.

9.—VIEW OF MESSRS. MAUDSLAY, SON, AND FIELD'S WORKS, LAMBETH.

Chapter XII.

THE ENGINEERING ESTABLISHMENTS.*

Maudslay, Sons, and Field.
The manufactory of Messrs. Maudslay, Sons, and Field is situated in the Westminster-road. It was first established towards the end of the last century in Margaret-street, Cavendish-square, and was removed to its present site in the year 1810. It covers several acres of ground, and is in most parts two, and in many three, stories high. It was in this manufactory that most of the earlier mechanical improvements were first introduced into general use, such as the slide-rest (that important adjunct to the lathe), the improved punching machines for boiler plates, and the machinery by which screw cutting was first reduced to a system. The possession of such tools naturally brought the best work to this manufactory, then conducted by the late Mr. Henry Maudslay, and amongst the earliest and most important works executed was the block machinery erected at Portsmouth Dockyard in the year 1804. These machines have supplied the whole British navy with blocks ever since (a period of sixty years). Although of late years marine engines have been the principal manufacture of Maudslay, Sons, and Field, yet many most important engineering works have

---

* This is not to be taken as a list of the engineering establishments, for these are very numerous.

been executed by them, such as the pumping engines for the Brentford, Chelsea, Lambeth, and West Middlesex Water Works; the machinery and pumps for the Commissioners of the Haddenham Drainage; and the engines and pumps for the Southampton and Copenhagen docks, for Sebastopol and Egypt. Much of the machinery at the Royal and Calcutta Mints was constructed at Lambeth, as also that for the Imperial Ottoman Mint; the Anglo-Mexican, the New Granada, and many other South American Mints; and the whole of the machinery at the great Government Tobacco Works at Lisbon, besides gun-boring machinery for Turkey and Brazil. Messrs. Maudslay also made the stationary engines by which the trains were drawn by rope from Euston-square to Camden-town, locomotives then stopping at the latter place; and they likewise made many locomotive engines for the London and Birmingham Railway Company. This led to the firm being employed on the large engines fixed at the Minories station for drawing the trains from Blackwall to the Minories terminus, rope traction being adopted for this purpose, as it had previously been from Euston-square to Camden-town. The engines for the atmospheric railway on the several stations from London to Croydon, and on the South Devon Line, were likewise made at Lambeth. In consequence of the quality of the work executed, Messrs. Maudslay, Sons, and Field have performed a great variety of smaller operations, which required not only good workmanship but much scientific knowledge. Amongst these may be mentioned the time balls at Greenwich, Edinburgh, the North Foreland, and Sydney, New South Wales, which are worked by electricity; preparing and finishing the

9.—VIEW OF MESSRS. MAUDSLAY, SON, AND FIELD'S WORKS, LAMBETH.

9.—VIEW OF MESSRS. MAUDSLAY, SON, AND FIELD'S WORKS, LAMBETH.

standard measures for the Government, and all the large work for the Equatorial at the Observatory at Liverpool. The great and most important operations of this firm have been marine engines, the first of which were fitted in the *Richmond*, a small vessel which was built in the year 1815 and plied between London and Richmond. This was followed in 1816 by the *Regent*, which ran from London to Margate, and many of the most improved arrangements have been patented by them. In 1827 the oscillating engine was patented, which was followed by the double cylinder (to avoid the wear and tear caused by cylinders of large power on the oscillating principle), the annular, the steeple, double piston-rod engine, which takes up less room than any other form of engine, and finally the three-cylinder economic engines. In 1825 the *Enterprise*, of 120-horse power, was fitted and in great part owned by this firm, and made the first voyage round the Cape to India, where she was immediately bought by the East India Company, and did very effective service on the Indian and Burmese rivers. On the completion of the Great Western Railway from London to Bristol Messrs. Maudslay, Sons, and Field were called upon to construct the engines for the *Great Western*, of 1,300 tons and 400-horse power. This vessel successfully performed the voyage from Bristol to New York for many years, and effectually solved the problem as to the possibility of steamships being able to cross the Atlantic, which had been denied by the late Dr. Lardner, at the British Association, held at Bristol in 1835. It is to the success of this vessel that we may be said to owe the formation of our present large sea-going mail packet companies. On the formation of the Royal Mail Steam-packet Company Messrs. Maudslay became shareholders, and made the engines for the mail steamers *Thames* and *Medway*, and subsequently fitted the company's ship *Orinoco* with engines of 800-horse power, which, after working many years, were taken out of this ship (which was broken up) and placed in their new vessel the *Paramatta*, which was unfortunately lost on her first outward voyage. The engines of her Majesty's first yacht, the *Victoria and Albert* (but now the *Osborne*), were made by this firm in 1842, and have worked and still continue to work with great efficiency. On the introduction of the screw propeller Messrs. Maudslay and Co. were intrusted by the British Admiralty with the construction of engines for her Majesty's ship *Rattler*, the first ship fitted with the screw in the Royal Navy. It was in this vessel that most of the experiments with the screw were made, and which led to its almost universal adoption. At this time Messrs. Maudslay, Sons, and Field built a small steam vessel called the *Water Lily*, in which at their own expense they made many experiments with various forms of screws, and obtained many results which have been useful as guides to themselves and others. This vessel was ultimately purchased by the Austrian Government and employed in the postal service on the Adriatic. Subsequently Maudslay and Co. constructed for the General Screw Steam Shipping Company a great many vessels of different powers, and fitted them with their patent feathering screws. This enabled the ships, most of which were about 1,800 tons burthen and 300-horse power, to take every advantage of their sailing qualities, and, as the pitch of the screw could be altered from the deck to suit the velocity of the vessel, and the blades even placed fore and aft, they possessed all the qualities of full-powered sailing vessels combined with steam. These ships ran from London to Australia for several

9.—VIEW OF MESSRS. MAUDSLAY, SON, AND FIELD'S WORKS, LAMBETH.

9.—VIEW OF MESSRS. MAUDSLAY, SON, AND FIELD'S WORKS, LAMBETH.

years, but from various reasons the company eventually broke up, and most of the ships passed into the hands of the London and East Indian Company, and have since been employed between London and Calcutta. The feathering screw has also been fitted to many gentlemen's yachts, and likewise to her Majesty's gunboat *Stork* and frigate *Aurora*, of 400-horse power. The public are greatly indebted to Maudslay and Co. for the first fast Channel boats, as in 1843 they, in conjunction with Mr. Mare, then of the Blackwall Ironworks, built and equipped the mail steamer *Princess Alice*, which vessel ran the distance in just half the time taken by the old packets. She was purchased by the Government within a fortnight after her being put upon the passage. This led to the present fast line of packets not only between Dover and Calais, but to the Channel Islands and between Holyhead and Dublin; and the firm constructed the *Princess Mary*, the *Princess Maude*, the *Queen of the French*, the *Queen of the Belgians*, the *Chemin de fer Belge*, the *Express*, the *Despatch*, the *Courier*, *Anglia*, *Scotia*, &c. The latest improvement which has been patented by this firm is the three-cylinder economic engine. This engine is fitted with every appliance tending to economise fuel, such as superheating, surface condensation, and using the steam expansively. Her Majesty's frigate *Octavia*, of 500-horse power, is fitted with these engines, and the saving in fuel was found to be just 50 per cent. in comparison with the usual consumption. The Russian iron-plated battery *Perrenetz*, of 300-horse power, is also fitted with engines of this description, as are also the London and Mediterranean Steam Navigation Company's vessels *Italia*, *Alexandra*, and *Clotilda*, which are running between London, Genoa, and Patras, and fully confirm all the experiments made in the *Octavia*. Although doing actual work at sea, the average consumption during the voyage with good coal is only $2\frac{1}{2}$lbs. per indicated horse-power per hour. The manufactory gives employment to from 1,000 to 1,500 men, and contains the tools and all appliances required for the manufacture of the largest marine engines. The various departments may be said to consist in the pattern-makers' shop, which is upwards of 200 feet long and 40 feet wide; two foundries (iron and brass), forge shops, two coppersmiths' shops, four vice-lofts, two boiler-makers' shops, turneries, punching shops, storerooms, and three large erecting shops, the largest of which is 150 feet long by 60 feet wide, besides large water-side premises. All the shops are provided with travelling cranes, which are capable of lifting from 25 to 30 tons, and run the whole length of the various shops. The tools consist of lathes (of which there are upwards of 80), 30 planing machines, shaping machines, rivetting machines, punching, drilling, screw-cutting, and facing machines, steam-hammers, boring machines, and crank-turning machines. These are driven by six stationary steam-engines. With regard to the capabilities of this establishment, it has on an average for many years past turned out marine engines of 5,000-horse power annually (exclusive of other work), and on emergencies much more, as during one year, at the time of the Russian war, this firm supplied the Government with engines for upwards of sixty gunboats, besides several pairs of larger engines. They have at the present moment the following engines either fixing or constructing:—

H. M. iron-plated frigate *Agincourt* .... 1,350 H. P.
    ,,   ,,   *Prince Consort*...... 1,000 ,,

8.—VIEW OF MESSRS. PENN AND SON'S WORKS GREENWICH AND DEPTFORD.

8.—VIEW OF MESSRS. PENN AND SON'S WORKS GREENWICH AND DEPTFORD.

H. M. iron-plated frigate *Ocean* .... 1,000 H. P.
,,   ,,   *Caledonia* .... 1,000 ,,
,,   ,,   *Valiant* .... 800 ,,
,,   ,,   *Zealous* .... 800 ,,
,,   ,,   *Royal Alfred* ...... 800 ,,
H.M. corvette *Harlequin* .... 200 ,,

Besides various marine engines for foreign Governments of the power of 4,000 horses; whilst up to the present time they have turned out, or are completing, marine machinery alone of an aggregate power of between 90,000 and 95,000 horse-power.

<small>Messrs. John Penn and Son.</small> Messrs. John Penn and Son's works are divided into two establishments, the engine works being at Greenwich, and the boiler manufactory at Deptford, on the banks of the Thames. Moored off the latter establishment is a hulk fitted with large shear-legs and cranes for the purpose of placing machinery of the heaviest description on board vessels. At the commencement of this century, Mr. John Penn, the father of the present eminent engineer, started a small millwright's and machinist's shop on the site of the extensive premises at Greenwich, and for some years devoted himself entirely to the manufacture of windmills, water-wheels, and the machinery connected with the same. Many of our large flour-mills and gunpowder works bear witness to his industry and skill, and the treadmills in most of our county gaols are some of his early works. About 1830 the son of this enterprising man directed his attention to the steam-engine as applied to mills and small vessels, and after a few years, not considering the beam engine suitable for marine purposes, he turned his thoughts to the oscillating engine, which is now so

intimately associated with his name The advantages of this style of engine are so universally acknowledged that it would be superfluous to describe them. On their first appearance they met with some opposition, but after several years' trial it was found that they occupied smaller space, weighed less, were simpler in construction, and were better suited for steam navigation at a high speed than the other engines in general use. This induced the Admiralty to order a pair of oscillating engines of 260-horse power for the paddle-wheel yacht *Black Eagle*, and their great success has caused the Government to adopt them in many ships of war, in her Majesty's yachts, and in most of the Royal Mail Packets. About the year 1840, when the Admiralty decided on using the screw as a means of propulsion for ships of war, the general arrangement and speed of paddle-wheel engines for driving the screw being found insufficient without the use of geared wheels, it was determined to seek designs from the most celebrated marine engine makers, with a view to construct a direct-acting engine suitable for the high speed required, all parts of which should be under the water-line. Most of the firms responded, and among them Messrs. Penn and Son, who brought out about this time the double trunk engine. The design was approved, and permission given to supply the vessels *Arrogant* and *Encounter* with engines of 360-horse power on that principle. Their successful performance led to the line-of-battle ship *Agamemnon*, 600-horse power, being similarly fitted, and up to this time the trunk engines have been applied to no less than 130 vessels in the Royal Navy, including the ironclad frigates *Warrior*, 1,250-horse power; *Black Prince*, 1,250-horse power; *Resistance*, and *Defence*, and are almost exclusively used for the Spanish and Italian

8.—VIEW OF MESSRS. PENN AND SON'S WORKS GREENWICH AND DEPTFORD.

8.—VIEW OF MESSRS. PENN AND SON'S WORKS GREENWICH AND DEPTFORD.

navies. It may be added that the engines for the new ironclad ships *Achilles*, 1,250-horse power; *Minotaur*, 1,350-horse power; and *Northumberland*, 1,350-horse power, are in course of construction at these works. Nearly 1,200 men are constantly employed in the various departments at Greenwich, and the works there, which cover about nine acres of ground, have within the last three years been greatly extended and partially reconstructed. The boiler manufactory at Deptford gives employment to about 600 men, and both establishments have been thoroughly furnished with the best modern tools, chiefly by Whitworth. Ordinarily, machinery and boilers for about 6,000 to 7,000 horse power are turned out annually, and in case of emergency this might easily be increased to 10,000 horse power. This alone would be sufficient for seven iron-cased ships of the largest size.

The Works. Altogether the works of the Messrs. Penn, like those of the Messrs. Maudslay, occupy in marine engineering much the same position as the Thames Company and the Millwall Company do in shipbuilding. They are the foremost in traditions, name, and magnitude, not merely in this country but in the world. France has no such engineering works, neither has America. Together they are equal to the whole marine engineering wants of the Royal Navy, in the greatest possible extremity. This is a fact, the suggestiveness and importance of which could scarcely be overrated. As may readily be supposed, the works of the Messrs. Penn are a model of economic adaptation and administrative skill. Complete efficiency is attained by vesting skilled foremen with a great amount of power and holding them responsible for its proper exercise. Into the hands of the foremen the work to be done is committed under the supervision of the members of the firm, and the foremen may on the instant discharge workmen for shortcomings or misconduct. The foremen, in point of fact, are the masters of the workmen, and, entering and leaving the works with the workmen, the latter are never at a loss; nor is there the opportunity, even if there were the will, to be idle or imperfect in the performance of appointed tasks. This is the secret of faultless engines; engines that will do their work without break-down, as long as coal and water are supplied, and indifferently in smooth and troubled water. Upon the best cast and forged metal the best workmanship is placed. And, to do Messrs. Penn's workmen justice, the yoke of supervision is an easy, nay, a pleasant one. Treated as men deserve always to be treated, the great body of them have attained that frame of mind which identifies their employers' interest with their own. An inquirer in Greenwich or Deptford will find that Penn's "men" are a class known to and esteemed by the whole community, chiefly because of the respect they manifest for themselves and of the esteem in which they hold the members of the firm. The distinguished foreign gentlemen who from time to time have taken off their coats in Messrs. Penn's works to become practically conversant with marine engine-building, no doubt look back with satisfaction on the time of intelligent association with Messrs. Penn's workmen. The superior administrative staff is small, which is another suggestive circumstance. Two or three able men, even in engine-building, easily direct a thousand. The argument is unanswerable against numerous staffs, and particularly where the work is less exact and skilled than in engine-building. It is also unanswerable against large outlays for mere control.

8.—VIEW OF MESSRS. PENN AND SON'S WORKS GREENWICH AND DEPTFORD.

8.—VIEW OF MESSRS. PENN AND SON'S WORKS GREENWICH AND DEPTFORD.

A detailed description of these great national works is not required. Suffice it to say that boiler shop, erecting shop, turnery, foundry, forge, &c., are unsurpassed.

The Albion Ironworks. The establishment of Messrs. George Rennie and Sons consists of two large factories, one in London, the other at Greenwich; the London works, in Holland-street, Blackfriars, running back to the River Thames, cover several acres. The workshops, of lofty and handsome elevation, supported on cast-iron columns, are replete with all the modern tools and appliances for the execution of mechanical and general engineering work on the largest scale, with a fine iron-foundry. Messrs. Rennie manufacture everything for themselves, and their boiler-yard has produced some of the largest wrought-iron boilers for the navy. The Greenwich branch is also of great extent, and fitted as completely as it can well be, for the building of iron ships, and some gigantic iron floating-docks for the Spanish Government were recently completed. The necessity of providing dock accommodation for the monster iron-clad vessels of war is so apparent that these floating-docks are entitled to consideration, especially for the colonies and our possessions in the West Indies and elsewhere. The name of Rennie is as distinguished in civil engineering and mill mechanism, as that of Watt in mechanical engineering. John Rennie and James Watt were men of immense talent, and both possessed indomitable energy. They were intimately associated in their lives, and their reputations are now to a considerable extent inseparable.

Rennie's history, and Progress of the Firm. Having finished his mechanical career in Scotland, and fairly established himself with Boulton and Watt, in England, Rennie repaired for the first time to London, in October, 1784. He established himself close to the Surrey side of Blackfriars-bridge, to be near the scene of his future labours—the new Albion Mills. The Albion Mills consisted of two engines of 50-horse power each, and twenty pairs of stones, of which latter twelve or more pairs were constantly kept at work. Rennie went on successfully with the mills until 1791, when a disastrous fire occurred, and in a short time destroyed the whole. The works were never re-erected. The walls still stand, however, and the interior having been converted into dwelling-houses, is now known as Albion-place, Blackfriars-road. Watt, in writing of the Albion Mills subsequently to their destruction, speaks of Rennie as "a valuable and able mechanician and engineer;" and states that the machinery erected by him "forms the commencement of a system of mill-work which has proved and will prove most useful to the country." Another and far more important achievement must be awarded to Rennie. The towing of the *Hastings*, 74, by a steamboat, suggested by him, led to the adoption of steam in the British navy, the *Lightning* being shortly after fitted up by Boulton and Watt, at the instance and under the superintendence of Rennie, for the Government. So early as 1801, Rennie reported upon the project of an iron rail or tramway between the east and west ends of London. He also reported, in 1810, on a railway from Berwick to Kelso. His last and most important report on railways, however, was on that proposed to be constructed between Stockton and Darlington, and which had been originally surveyed by

10.—VIEW OF MESSRS. RENNIE'S WORKS, DEPTFORD AND BLACKFRIARS.

Messrs. Brindley and Whitworth in 1768. These reports are elaborate and valuable documents, and contain minute and exact comparisons between the cost of canals and railways. The adoption of the railway at Stockton followed Rennie's favourable report, and the gradual substitution of the locomotive engine—then a rude machine in the hands of Blenkinsop—by George Stephenson originated the present railway system. It may be added that the surveying and laying down of the first passenger and goods railway in England—that between Manchester and Liverpool—was effected by the two sons of Rennie, George and John, the first-named acting for both in carrying the Bill through Parliament. In 1836 Messrs. Rennie were called on to remodel the mechanical and engraving departments of the Bank of England. This was accomplished under the superintendence of Mr. Oldham, of the Bank of Ireland, and the result was eminently successful. From 1837 to 1842 the Holland-street establishment was much engaged with the construction of locomotive engines, furnishing several for Cuba, the Austrian and Italian lines, and those of Brighton and Croydon. It was found that considerable inconvenience arose from the prosecution of this branch of manufacture—the premises not being well adapted for carrying it on—and it was abandoned. About the same period the Russian Government, which was fitting up a large naval arsenal at Nicholaeff, in the province of Kherson, and on the shore of the Black Sea, anxious to avail itself of all the improvements which had been introduced into the dockyards of England, determined on being furnished with a set of block machinery similar to that at Portsmouth. Messrs. Rennie were applied to, and they undertook its construction. The funds to be devoted to the purpose

T

were, however, less ample than could be desired, and the engineers were obliged to reduce the scale of the machinery. Perhaps one of the most extraordinary works in the shape of dock engineering effected by Messrs. Rennie, or by any other English or foreign engineers, were the massive iron gates supplied to the docks of Sebastopol. These were of dimensions far beyond anything of the kind made before or since, and a leviathan planing-machine, expressly for fitting them up, was contrived by Mr. G. Rennie, and erected in the Holland-street factory. These gates, which contained many hundreds of tons of wrought and cast iron, were successfully transferred to the once great Russian stronghold of the Black Sea, erected there, and partially if not entirely destroyed by the cannon of England and France during the siege of Sebastopol. Another work of importance executed by Messrs. Rennie was the machinery for the Turkish Small Armoury, situated on the Sweet Waters near Constantinople. It was designed to manufacture 100,000 muskets per annum, and it comprises the whole of the appliances for forging, rolling, rough and fine boring, adjusting, and proving the barrels; the forging and fitting of the different parts of the locks; the cutting and shaping of the stocks, &c. There were, on the whole, eight sets of rough boring machines, four sets of fine ditto, and twelve turning lathes of peculiar construction for giving proper taper to the barrels. The whole was to be driven by a 30-horse power engine. The plan for the buildings to receive this machinery, and a massive rolling-mill for shaping the skelps of iron intended for barrels in addition, was given by the firm, and it was of a most comprehensive and complete character. The Armoury at Constantinople has been in successful

10.—VIEW OF MESSRS. RENNIE'S WORKS, DEPTFORD AND BLACKFRIARS.

operation ever since, and during the Crimean war was worked to its utmost capability. Marine and land engines were also constructed for the home and foreign Governments. Among the first of these may be mentioned a pair of 300-horse power for the *Archimedes*, a screw steamship built for the Russian Government, and the first introduced into the Russian navy. The celebrated *Vladimir*, so well known during the Crimean war, a paddle-wheel boat, and remarkably swift, was fitted up by the Rennies. Many others of less power, some propelled by screws, and others by paddle-wheels, were also furnished to the Emperor Nicholas for service in the Baltic and Black Sea. For her Majesty's Government they have supplied engines for the *Bulldog*, 500-horse power; the *Vulcan*, 350-horse power; for the *Oberon*, the *Melbourne, Samson, Reynard, Courier*, and forty or fifty others.

The Floating Docks. The mode of docking ships in Rennie's patent floating docks is somewhat similar to that employed in the ordinary process with other docks. The dock is submerged to the extent required by the draught of water of the vessel to be repaired. The latter is then to be warped in, and placed over the keel blocks. The water is then pumped from the compartments of the docks, and the whole mass is gradually raised, the vessel being shored from the altar course and made secure on her seat. Very soon the ship becomes high and dry, although of 7,000 tons, and prepared for examination and repair. When necessary to undock the ship, nothing more, of course, is required than for the sea sluices to be opened, and for water to displace the air, which has outlets formed in the decks of the pontoons from the compartments. Then the vessel floats, is warped out, and the dock becomes ready for another tenant. One remarkable characteristic of the floating docks is, that they are quite open at each end. There are no gates, caissons, or any means of closing them, and thus the greater length of a vessel will be no bar to her entrance. Weight is the only point to be considered. The Carthagena Dock, which has been completed, is 320 feet in length, 105 feet in breadth over all, and the depth of its base is 11 feet 6 inches. The clear space within its walls is 78 feet 6 inches, and the inner depth is about 45 feet. The Ferrol Dock is 350 feet in length, 105 feet in breadth, and the depth of its base will be 12 feet 6 inches. In other respects its dimensions accord nearly with those of the Carthagena dock, and the quantity of material used in both is nearly equal. To the consideration of these docks the attention of the Admiralty cannot be too strongly urged.

The Phœnix Foundry, Liverpool. Among the leading engineering works in the town of Liverpool are those of Messrs. Fawcett, Preston, and Co., and so long have they been established that at one period there was no other in the town. Some eighty years ago the foundry was not in existence; but in small premises at the corner of York-street the Coalbrookdale Foundry (in Shropshire) had a depôt, managed by Mr. Rathbone, the maternal uncle of the late Mr. Fawcett. On the death of Mr. Rathbone Mr. Fawcett became his successor. Mr. Fawcett during the greater part of his life resided in the old house at the corner of Lydia Ann-street, now a portion of the works; he was born in 1761, and died in 1844, in the eighty-third year of his age. During the long war, a great demand having arisen for iron guns to enable merchant ships to cope with privateers and vessels sailing under letters of marque, Mr. Fawcett introduced the highly important improvement of casting the guns solid and boring them, as is the practice still, the previous custom, however, having been to cast them hollow. Mr. Fawcett's innovation was so favourably received by the public that the annual value of the guns turned out by him was not less than £10,000—a very large sum for such a purpose in those days. The carronades marked "solid" on one trunnion and "F" on the other may be seen now all over the world, and their origin identified.

<small>Progress of the establishment. Guns.</small> The establishment increased and prospered, shop after shop being added to it, with appliance on appliance, until the present large proportions have been reached. In addition to the foundry, there are now the brass foundry in York-street, and the extensive boiler yard and copper shop in Lightbody-street. Castings of the heaviest description are run in the foundry, among which are anvil-blocks for the Mersey forge, the heaviest weighing no less than 62 tons, and standing on a base of 110 square feet, foundation-plates for the cranes, and the fly-wheel weighing 60 tons. At this foundry the manufacture of guns is carried on extensively,—not the simple ship-guns of former days, but every description of improved artillery, from the light steel mountain gun on its wrought-iron carriage to the heavy-built rifled Blakely 300-pounder. At present Messrs. Fawcett, Preston, and Co. have several 9-inch guns in various stages of manufacture, as well as numerous others of all classes. To the Peruvian and other foreign Governments they have supplied large quantities; and during the Crimean war our own Government was supplied by the firm with

a large quantity of heavy sea-service mortars, weighing each 5 tons.

<small>Land and Marine Engines, &c.</small> Beyond the foundry, there are various workshops for the manufacture of the largest land and marine engines, hydraulic and other presses, rice and sawing machinery, water-wheels, caloric engines, and all the varieties of sugar apparatus for the refineries of Cuba, Java, the Mauritius and the world, from the little cattle cane mill up to that with 7-feet rolls, each weighing as many tons, and from the open iron teache to the largest vacuum apparatus. In fact, the manufacture of sugar apparatus may be said to be a *spécialité* of the firm, and Liverpool will say that their reputation has been earned deservedly. Of marine engines the firm have made many. Among other vessels of mark, they fitted the *Leeds* with engines in 1826. This vessel was built for the City of Dublin Steampacket Company, to ply between Dublin and Bordeaux, and her performances so well pleased the French Government that they ordered several pairs of engines for war steamers, the *Sphinx*, the *Gomer*, the *Asmodee*, and others, the two last of 450-horse power. The *Messageries Impériales* have also recently supplied a large portion of their fleet with engines made by this firm. The firm also made the engines for the following well-known ships:—The *Quorra*, the first iron steamer ever built; the engines for the ill-fated *President*; the engines for the *Royal William*, the first steamer that ever crossed the Atlantic from Liverpool to New York; the engines for the *Tigris*, the *Euphrates*; the engines for the *Merlin*, *Medusa*, *Medina*, mail steamers for the City of Dublin Company; and a whole fleet for the Peninsular and Oriental Steam Navigation Company—the *Oriental*,

Hindostan, Bentinck, Orissa, Behar, Ottawa, Malta, Nubia, and *Alma;* and last, but not least, the engines for her Majesty's frigates *Inflexible, Resolute,* and *Assistance:* the engines for the *Ganges* and *Jumna,* for the Oriental and Inland Steam Navigation Company, are also the work of the firm, and at present several pairs of very handsome engines of high power are in hand.

Contract Steamers. In addition to fitting engines for others, they also contract for steamers complete, and of those they have already supplied are the *San Luis, Pindari, Itapicuni, Capias,* and *Camossim,* for the Brazils. The *Oreto,* for Palermo, is another of the steamers contracted for by this firm, and it is scarcely necessary to remark that she has since changed hands and is now sailing under the flag of the Confederate States as the *Florida.* There are still others: the *Phantom,* a steel steamer of great speed and beauty, which, by recent accounts, had found its way into Wilmington, N.C.; the *Alexandra,* still detained by Government until the legality of her construction is tested on appeal; the *Great Victoria,* for the Australian line; and also two beautiful and swift steel steamers on the stocks. Of steam tugs, dredge boats, and barges it is unnecessary to speak, these having been supplied in great numbers. Altogether, of the capabilities of the engineering works of Messrs. Fawcett, Preston, and Co., it is impossible to convey an impression. The machinery for boring, turning, and other purposes, is on a scale of the first importance, and the last order for engines on the books is 2,307. Rightly has the establishment been called the Phœnix Foundry, for in 1843 it was burnt to the ground, which, although a serious injury at the time, ultimately proved an advantage, as in rebuilding the premises they were much improved in arrangement and convenience, and the antiquated tools replaced by others of the latest and most effective description.

<small>The Blackwall Ironworks. Stewart.</small> The Blackwall Ironworks have been in existence nearly twenty years, during which time, by talent and perseverance, the business has steadily increased. Mr. Stewart originally commenced business as a manufacturer of engines for tugboats—a class of machinery requiring peculiar construction and great strength, inasmuch as it is subjected to various and severe strains when towing vessels of large tonnage in a heavy sea; and in addition to the greater number of the best tugs on the Thames, he has manufactured similar engines for France, America, India, and Australia. Nearly all these tugs have paddle-wheels fitted with feathering floats, some having bearings on the floats, where wood and brass are made to work together; and the result of this combination is so favourable, that wheels so constructed have been working nearly six years with so little wear that they may go another six years without the bearings requiring to be renewed. Mr. Stewart has now extensive river-side premises, having built large shops, well supplied with machinery and plant of the best description, by which means he has been enabled to manufacture marine engines and boilers of a large size, and has fitted them to first-class mercantile vessels with the most satisfactory results. Amongst these are two fine steamers for the Australian passenger trade, one of 220-horse power, the other 150-horse power; the former was built by the Thames Iron Shipbuilding Company, from designs by Mr. James Ash, the latter by Mr. Charles Lungley of Deptford; also three steamers to run between Southampton and the Isle

5.—VIEW OF MR. STEWART'S BLACKWALL IRONWORKS, MILLWALL.

5.—VIEW OF MR. STEWART'S BLACKWALL IRONWORKS, MILLWALL.

5.—VIEW OF MR. STEWART'S BLACKWALL IRONWORKS, MILLWALL.

of Wight, well known for their speed and efficiency. At present there are in hand three pair of oscillating engines of 150-horse power each, for the Australian passenger trade; a pair of oscillating engines of 225-horse power, for the South-Western Railway Company, to be fitted with surface condensers, and intended to run between Southampton and the Channel Islands, have just left the factory; a pair of engines of 250-horse power for the Chinese trade; also eight pair of towing engines of various power, the whole to be fitted with all the latest improvements. Besides these, orders have been received to design and build a pair of screw engines of 60-horse power, to be fitted with surface condensers, for a yacht for Arthur Anderson, Esq., chairman of the Peninsular and Oriental Steam Navigation Company; these engines are to be in every respect the best that can be made. The hull, &c., is being constructed by Messrs. James Ash and Co., the dimensions being as follow:—Length between perpendiculars, 144 feet; breadth, 22 feet; depth, 12 feet. Her tonnage is 337 tons, and she is to be fitted in the most sumptuous style.

Engine and boiler works. The engine and boiler works are situated on the Isle of Dogs, near Messrs. Samuda's shipbuilding yard, and although not occupying so large an area as some other factories, possess the advantage of unusual compactness and economy of space. The river frontage is 400 feet in length, and is so arranged that there is a kind of wet dock, over which two powerful travelling cranes are made to traverse, thus affording remarkable facilities for putting boilers and other heavy parts of machinery into steamers. That portion of the factory near the river is devoted almost exclusively to boiler-building, and has a fine shop provided with every necessary and convenience for the purpose. At the back of this are built the various shops for the manufacture of steam-engines—viz., fitting and erecting shops, pattern shop, blacksmiths' and coppersmiths' shops, &c., all of which are built from Mr. Stewart's own designs, and are in all respects first-rate. A large new foundry is now in course of construction on a piece of ground adjoining the factory, and when this is completed the Blackwall Ironworks will in all respects be most complete. Altogether the Blackwall Ironworks furnish another proof, were one required, of the great superiority of private enterprise and the contract system over the wretched jobbery, inefficiency, and unproductiveness that prevail in the dockyards. Organisation, industry, and economy are visible at every turn. There is not an idler to be seen, and none of those tell-tale stacks of abused material that disgrace the dockyards. Successful private enterprise is incompatible with dead weight of any kind. Its conditions are clear-headed calculations, careful working up of raw material, and a good understanding alike with the skilled and unskilled workmen. Mr. Stewart enjoys the zealous and invaluable co-operation of his sons. His business grows upon his hands, and his means of efficient management keep pace with the increase. The same might be said of many other successful firms; but there is no such thing in the dockyards. Every extension of dockyard work implies an increased area of waste and unproductiveness. If already the dockyard system is not insupportable to the taxpayer, the new works at Chatham, Portsmouth, and Keyham must make it so; while, on the other hand, the development of private industry is at all times an honour and advantage to the country.

7.—VIEW OF THE BOATBUILDING COMPANY'S WORKS, EAST GREENWICH.

Chapter XIII.

SHIPBUILDING INNOVATIONS.

The National Company for Boatbuilding by Machinery (Limited). Foremost among the shipbuilding innovations of the time is that of boatbuilding by machinery. The application is most successful, and in its success there is obviously a strong argument against the maintenance of all the minor dockyard manufacturing establishments, because improvements in boatbuilding imply the possibility of improvements in all other directions—nay, the certainty of such improvements, were the Admiralty entering the market in a straightforward business manner for the supply of all their minor wants, in ropes, sails, blocks, &c. Mr. Nathan Thompson, the inventor of the improved boatbuilding machinery, luckily for himself and the public, did not share the fate of most inventors, but, under the auspices of such men as Colonel Sykes, Peter Graham, and John Dillon, succeeded first in demonstrating the utility of his "tools," and afterwards in establishing a company, which, enjoying the entire confidence of the mercantile marine, is barely able to keep pace with the increasing demand for boats. The company have had, and may still expect to meet, many difficulties, and to overcome many prejudices;—doubts as to the quality and durability of the boats turned out by them—ignorance, and in many cases worse than ignorance, on the part of the men employed on the machines—the interests of competing boatbuilders—the opposition to everything that is new. These, with threatened strikes and many other difficulties, have been one by one overcome, and, under the guidance of its directors and of its manager, there can now be no question as to the future which is before this company. The quality of every description of work turned out by the company is now so universally admitted to be superior to everything previously seen, that from boats suited to the mercantile marine the directors have been compelled to enter upon the erection of barges, canal boats, and vessels of large size. To accomplish this they are now enlarging their premises to an extent that a year or two ago even Mr. Thompson perhaps did not think probable. Boatbuilding in point of time is now reduced from a matter of weeks to a matter of hours, and, as regards expense in the matter of labour, is reduced from pounds to shillings; the cost of material, of course, being the same whether boats are built by machinery or manual labour. The company's premises are situated at East Greenwich, about half a mile below Greenwich Hospital, and cover an area of nearly nine acres of ground. Every possible appliance to lessen the cost of production has been provided—wharves, steam-cranes, tramways, &c. The number of men employed under the able superintendence of the Messrs. Fawcett, formerly of Limehouse, and temporarily of Messrs. Nathan Thompson and John C. Thompson, is about four hundred, and there is perhaps no place in the neighbourhood of London more worthy of a visit than the works of the company. Orders to respectable parties are readily given on application to Mr. Grant, the company's secretary, at the London office, 123, Fen-

7.—VIEW OF THE BOATBUILDING COMPANY'S WORKS, EAST GREENWICH

church-street. When the scheme of Mr. Nathan Thompson was first submitted to the public, the writer supported it in the following terms:—

Machine Boatbuilding.* "When New England ingeniously and unscrupulously began the manufacture of wooden nutmegs, it was apparent that there was no conceivable limit to the application of machinery and steam. To counterfeit the genuine article which housewives use so freely, by giving an irregular oval form to little knobs of New England mahogany, was to let daylight into the economy of trade, and to suggest no end of refined deceits. Wooden cheese and butter for the windows of provision shops, and wooden sugar-loaves for grocers' shelves, if not for their drawers and hogsheads, were no doubt among the earlier imitative efforts; and of the ultimate development of the science we of course cannot speak. But if the manufacture of nutmegs was suggestive of objectionable imitations, we have only to turn to the working of the patents of that eminent scientific American, Mr. Nathan Thompson, jun., to be assured that if inventive genius is stimulated in a bad direction, it is sure speedily to be stimulated in a way that is not only unobjectionable, but useful. We do not mean to say that machine boatbuilding is the offset to any invention that has preceded it, or that it has been suggested by any previous invention, but what we are sincerely anxious about is that our readers should be impressed with the fact that a revolution is impending in the boatbuilding business; that Mr. Thompson's machinery is eminently practical and singularly expeditious and economical; that the work it turns out is

---

* Article furnished by the writer and published in the *Steam Shipping Chronicle*, 21st June, 1861.

more exact and perfect, and, as a matter of course, stronger than the boats built hitherto by manual labour. Mr. Thompson's boatbuilding by steam is not for a moment to be thought of in connection with the manufacture of nutmegs or with the many spurious American inventions which by some means or other are introduced in England; but it is an intelligent and intelligible fashioning of the timbers and parts of boats and vessels in an expeditious manner, and so fashioning them that the vessels so constructed may be taken to pieces and fitted up at pleasure with little or no trouble.

<small>Advantages of Machine-made Boats.</small> "The advantages of taking boats to pieces are obvious. Why our emigrant, and troop, and other ships have usually an insufficient number of boats to carry all hands is owing chiefly to the space taken up by boats on deck, which involves interference with the health and comfort of every one on board. For the want, therefore, of such an invention as Mr. Thompson's the number of boats carried has generally been inadequate, while no provision whatever has existed for the loss of boats from swamping or from being stove in. On an emergency, when boat after boat has been got over the ship's side, and almost immediately afterwards turned over, with the unhappy creatures in it who hoped for safety, no boat whatever has, frequently, remained for the great majority of the crew and passengers. By the use of Mr. Thompson's boats a reserve of boats may be kept below, not only sufficient to take off all hands, but to provide against the casualties of launching. Then it is well known that boats hanging from the davits, or otherwise exposed on deck, suffer greatly from exposure, particularly in the tropics, and it will, therefore, be found economical in many cases to

have most of our southern-going ships' boats stowed away. Lastly, duplicates of any part of a boat may be supplied, and instead of boatbuilders' accounts for repairs, the duplicate can be inserted without any difficulty by any one. Thompson's boats, in short, go together like a bedstead, and it is only necessary to know where the different parts should go, to rig out anything, up to a pleasure yacht of one hundred tons burden.

<small>Utility in the Navy.</small> "To the Navy the invention can hardly fail to prove invaluable. Boats sufficient to land an army may now be stowed away in a single transport, without, as heretofore, inconveniently encumbering the decks of the transports or the ships of the covering fleet; and boats which may be taken down and packed up will be available for interior transport when ordinary boats would be of no use whatever. Africa, India, China, and even North America (Mr. Thompson's own country) are suggestive fields for the employment of boats which might be carried overland before being launched upon their proper element. And rifled cannon, it is scarcely necessary to observe, threaten to be most destructive to the old-fashioned boats of our ships of war. An action, now-a-days, at close quarters, will, if it does not lead to the annihilation of the ships engaged, render every old-fashioned boat that is exposed entirely useless, while Mr. Thompson's boats would come out of action all but scathless. Half a dozen round shot passing through them would only lead to the unshipping of the shattered fragments and to the fitting-in of duplicates, or to the repairing of one boat in an hour or two with the vestiges of another. So long as we were without an invention of this kind, the seamen in our

ships of war were, in fact, unsafe; and now that this is admitted, we trust that Mr. Thompson will not be treated by the Admiralty as Mr. Trotman has been. We hope to see the new machinery in operation in all our dockyards before the year is out, that the service may profit by increased efficiency, and the votes of next year be reduced by the economy which is sure to follow.

<small>Success predicted.</small> "Whether it is the intention of Mr. Thompson to grant licences for the working of his patents we, of course, cannot tell, but at the moment it is in contemplation to establish a joint-stock company with sufficient capital to supply a fourth of the 25,000 boats wanted in the United Kingdom annually. That such a company will be formed we cannot for a moment doubt, and that it will succeed is a matter upon which, in scientific circles, no doubt whatever appears to be entertained. The 'innovation' is looked upon as possessing much the same recommendations for boatbuilding, as the improved frame does for the spinning of yarn, and the improved loom for the weaving of textile fabrics. It is a shorter and cheaper and better way of arriving at a given result, as railway or steamboat travelling is the shortest, cheapest, and best way of getting to a journey's end. It is, in fine, a great mechanical step forward, and those who are wise will accept it as such. Boatbuilding by manual labour is about to be numbered among the things that were; boats will be produced cheaper than they yet have been; will be wanted for purposes to which hitherto they have not been applied; and although a considerable present displacement of labour will unquestionably be occasioned, all experience shows that eventually a greater number will earn their bread by building boats

than do just now. For one who made a living a hundred years ago by the spinning-wheel, ten thousand, if not ten times ten thousand, are now constantly and remuneratively employed.

Machinery employed. "We cannot, perhaps, do better than close these remarks with an account of the number and purposes of the machines employed by Mr. Thompson in the manufacture of a boat thirty feet long in a few hours. The first machine is called the 'assembling form,' which is for holding the gunwales, risings, floor-timbers, cants, keels, stem, stern-post and board in their relative positions, as designed in the finished boat. The second is the combination saw, for all kinds and dimensions of stuff, either square, bevelling, or angling, that can be sawed with a circular saw, and to any desired width or taper without measuring. The third is the patent form for spiling, or giving the plank edge the required bevel throughout its entire length. The fourth is for giving the proper bevel to the stern-board, thwart-knees, transom-knees, breast-hooks, risings, forward and stern-ribs, cants, stern-sheets, gratings, toggels, &c. The fifth is for bearding and rebating keels at a single operation, and in the most perfect manner. The sixth machine is for tenoning toggels. The seventh for marking and slotting gunwales to receive their toggels and rowlocks. The eighth for grooving, grating, &c. The ninth for giving the ribs their required bevel. The tenth for planing a plank on both sides at one operation, at the same time giving its interior and exterior curve in the most perfect manner, and uniform in thickness throughout its entire length. The eleventh is a machine for planing perfectly plane surfaces. The twelfth is for moulding toggels, bottom boards, gunwales, and risers; and it cuts any bevel or irregular mould, or three sides, or planes three flat surfaces at a single operation. The thirteenth and last machine is for bending the ribs to any form or size required in boatbuilding."

The Tripod Masts. The tripod masts, the invention of Captain Cowper P. Coles, is another suggestive innovation. These iron masts should supersede dockyard mastponds and mast-houses, and, to a very great extent, dockyard rigging manufacture and rigging-houses. They consist of three-legged iron masts, the upright centre leg being the mast proper, and the other legs the supports in place of rigging. A more admirable invention to promote efficiency and save money can scarcely be conceived. But it remains in the category of untried or rather unapplied inventions. To adopt it would be revolutionary, and its adoption will, therefore, be as long as possible delayed. Captain Coles claims for his tripod masts the following among other advantages*:—

1. From there being no lower stays, the yards can be braced nearly fore and aft, or in a line with the keel, enabling a ship so rigged to set her square sails when under steam, when with the present rig she would be obliged to keep them furled. On steaming head to wind, the yards being braced fore and aft, there being little rigging or top hamper on the topmast, there would be but a small resistance to the wind.

2. When these masts are shot away in action they would, being of iron, sink at once, as would the yards from the same cause. This, with the absence of lower rigging, would greatly lessen the chance of the screw

---

* Lecture delivered at the Royal United Service Institution, 25th March, 1863.

being fouled, one of the greatest dangers to which our ships are exposed.

3. The last, though not the least, advantage to be derived from the adoption of the tripod masts would be the saving which would be effected in the expense annually incurred through the enormous deterioration in the rigging of those men-of-war which it is thought necessary, according to the present system, to keep ready for service in the steam reserve. Ships with these tripod masts would always be comparatively ready for sea, and whilst their perishable sails and running rigging remained stowed away below, nothing would be exposed to the mercy of the elements but their iron masts and yards. The men go aloft by Jacob's ladders.

<small>Submarine Batteries.</small> It is well known to the Admiralty that scientific officers of the Navy are likely before long to perfect submarine batteries. These batteries contemplate firing the heaviest ordnance from the submerged bottom of one ship at the submerged bottoms of other ships. Mere mechanical difficulties, not by any means deemed insuperable, stand in the way of this being done; and the thing accomplished, another reconstruction of the Navy appears inevitable. But it will be answered, let us think of this invention when perfected, and in the meantime proceed as we are doing. It must be answered, no. On the contrary, let us rather proceed on the assumption that in time submarine batteries may form the most destructive armament of ironclads. America every day teaches us important lessons, and among others the one that explosions are as practicable under the water as out of it. Why, then, proceed on the principle that if above water ships are practically invul-

nerable, they may be as slim as you like below the armour shelf downwards? No doubt the wise policy would be to stop new ironclad construction altogether for a time, and in the meanwhile push forward the conversion of wooden ships, obtaining high speed by the substitution of engines of great power for engines of moderate power. This, of course, would compel us to resort to cupola instead of broadside ships; but who will show that a country possessing a cupola fleet is not as ready and as formidable as a country possessing a broadside fleet?

<small>The *Connector* experimental Ship.</small> The *Connector* experimental ship is the last innovation to which attention need be called. This is a ship constructed on the principle of a railway train; in other words, on the principle of detached sections. There is the motive section with the engine, boilers, and coal bunkers, and there are as many more readily attached and detached sections as you please. Many people laugh at the contrivance, but that has been the fortune of most things. In the *Connector* there is much for reflection, and this among other points, that in one or more of the attached sections you will always find a comparatively steady gun platform, no matter how it blows or how troubled the water may be. This has been proved experimentally by repeated voyages from the Tyne to London in heavy gales. It is, however, no part of the duty of the Admiralty as at present constituted to trouble themselves about the *Connector*, submarine batteries, tripod masts, or anything whatever out of the official beaten track. The Admiralty as at present constituted are worried to death with inventors and innovators. But

they are constitutionally opposed to all. They are themselves an institution of the time of Henry VIII. of happy memory, and their mission was, and still is, to glorify wooden walls, pigtails, and cat-o'-nine-tails. Henry VIII., in his most hopeful moods, surely never counted on the England of the present time being positively cursed with an establishment of his contriving.

Chapter XIV.

THE THAMES SHIPPING INTERESTS.

Deptford-green Dock-yard.† Mr. Charles Lungley's yard at Deptford-green is without exception the most complete on the Thames, combining as it does shipbuilding and repairing and the manufacture of steam-engines. It is on the Thames what the yard of Messrs. Laird is on the Mersey, with this difference, that the Messrs. Laird have spent large sums in the finishing of

---

* This is not to be taken as a list of the Thames shipbuilding firms, for these are very numerous.

† Mr. Lungley is the patentee of an important mode of interior fitting known as unsinkable shipbuilding. It will be generally interesting to give an extract from the provisional specification:—

"The object of the first part of my invention is to construct iron ships and other vessels in such manner that they shall not be liable to be sunk by any of the casualties which occur to vessels on the open sea, or by striking upon rocks, or by driving upon a lee shore, while at the same time their strength shall be increased and their accommodation for the stowage and carriage of cargo shall not be interfered with, as it now is where numerous transverse bulkheads are introduced into ships for a like purpose.

"The primary feature of this part of my invention consists in dividing the lower part of the ship or vessel into two or more water-tight compartments, and in affording access to these compartments for the introduction of cargo or stores by means of water-tight trunks or passages led up from them to such a height that their upper or open ends shall never in any practicable position of the ship be brought quite down to the level of the water; compartments thus formed may be used as ordinary cargo spaces, store rooms, chain lockers, or for any other like purposes, and may be ventilated by suitable trunks or tubes, always providing that all trunks or tubes of every kind which enter them shall be made water-tight and shall rise to the height before mentioned, in

6.—VIEW OF MR. CHARLES LUNGLEY'S DEPTFORD GREEN DOCKYARD.

6.—VIEW OF MR. CHARLES LUNGLEY'S DEPTFORD GREEN DOCKYARD.

docks and buildings. This Mr. Lungley has not done, and the liberties that he takes with his one dock are suggestive and remarkable. The dock entrance is a caisson fixture wide enough to take in the largest class

order that if by any mischance any compartment should be broken into, and the sea be admitted to it, the water should have no means of escaping therefrom into any other part of the ship.

"In carrying this part of my invention into effect, I vary the mode of applying it according to the form of the vessel and the service for which she is to be employed. In the case of a steamship for carrying both passengers and cargo I prefer to construct an internal bottom or deck in water-tight connection with the sides of the ship, and extending (where the arrangements of the boilers and engines will admit of it) quite fore and aft, at a height of several feet from the outer bottom or bottom proper. The compartment thus formed in the bottom of the ship may be divided transversely, if desired, by bulkheads, such bulkheads extending either to the top of the compartment only, or to any greater height as may be desired. Above this lower compartment, or set of compartments, and along the sides of the ship, I build vertical or inclined bulkheads, forming other water-tight longitudinal compartments, which again may be subdivided transversely, and which also are entered and ventilated by trunks or passages rising to the height before mentioned. With these arrangements it is evident that any portion of the submerged skin of the ship may be stove in by collision with another ship, or be torn away by rocks or otherwise, without causing the ship to sink. Supposing the remaining internal water-tight portions to be, as I always make them, of sufficient capacity to keep the ship buoyant and seaworthy, I sometimes form apertures in the inner bottom or deck for the purpose of letting any water that may get into the ship from above run down into the bottom; but these apertures are closed by valves or doors which are never opened except for this purpose, and are closed directly the letting through the water is completed: vessels built with these my improvements will not, therefore, be liable to those accidents which occur in ships fitted with water-tight compartments in the ordinary manner, and which result from passages through the bulkheads being formed and left open. The space occupied by the engines and boilers of a steam-vessel I close entirely in by water-tight iron walls or bulkheads extending to the same height above the water line as the trunks before referred to, in order that this space may be converted into a water-tight compartment from which water could not escape into any other part of the ship, and into which water could not enter from any other part. Apertures are formed in these walls or bulkheads for the admission of coals from the coal-bunkers, but these apertures are provided with valves or doors which may be closed either from below or from above. I sometimes further divide the boiler-room from the engine-room by a transverse bulkhead, in order that, should the engine break down or the engine-room become flooded, the boilers may still be kept at work and the steam be used to work pumps by means of an auxiliary engine in the boiler-room. I form divisions by bulkheads across the ship above the skins, which I term 'between-deck bulkheads,' and which also are made perfectly water-tight, and so as to divide the between-deck space in such manner that should the vessel ship seas, or otherwise get water on board, it may be confined to the part where it enters."

of ship or steamer trading to the Thames; but the dock itself appears to be diminished and enlarged by turns, to suit the varying wants of commerce. Frequently as many as three large vessels may be found together in the dock, and depth or width is provided by labourers with the spade. Occasionally, again, it may be found that the capacity of the dock is interfered with by the laying down of new ships in what seem out-of-the-way places which either encroach on the dock or which launch into the dock. Such is the fertility of private enterprise. Such are the expedients that successful private individuals resort to. Why not have similar economical docks in our remote possessions for the repair of ships of war? They might not be very sightly in the dockyards, but they would be quite as serviceable as any that can be provided. Mr. Lungley would probably require short notice to dock and repair the *Warrior*. Those, therefore, who are always deploring the want of docks overlook the fact that docks to any required extent may always be provided at a few weeks' notice. The Admiralty have only to invite tenders for the overhaul of the *Warrior* and the other ships to receive the offer of more private docks than they would care to use.

The capabilities of Mr. Lungley's Yard.   The capabilities of Mr. Lungley's yard are very great. The yard is well stocked with machinery, and constantly receiving additions to meet the requirements of the present busy time. During the Russian war Mr. Lungley built several vessels for the Admiralty, not one of which has been condemned or broken up,—a fact alike creditable to the builder, and to Mr. Letty, the dockyard inspecting officer. Several of the steam-vessels of the fleet of the

3.—VIEW OF MESSRS. SAMUDA'S YARD, MILLWALL.

Union or Cape Mail Company have been built by Mr. Lungley,—two of the number, the *Briton* and the *Roman*, being partially fitted on Mr. Lungley's unsinkable system. For the Australian trade Mr. Lungley has also built several most successful light draught of water steamers, and recently he has turned out some first-class iron sailing ships for the China trade, the property of Messrs. Phillips, Shaw, and Lowther. Mr. Phillips, the senior partner of that firm, is, it is well known, the deputy chairman of Lloyds' Committee. But one of the most creditable pieces of workmanship from Deptford-green Dockyard was the *Lancashire Witch*. This was an American-built ship, put into Mr. Lungley's hands for strengthening and repairs; and so well was the work performed that Lloyds' Committee adopted the *Lancashire Witch* as the type of repairs for the enlargement of their rules for the restoration of ships to the first class for lengthened periods. This occurred a few months ago; and Lloyds' Liverpool Committee visited the *Lancashire Witch* before the vessel was floated out of dock, expressing with the London Committee their high approval of the new plan of strengthening, and the superiority of the workmanship.

Messrs. Samuda's Yard, Millwall.* Messrs. Samuda's yard is one of the best-known on the Thames, and has

---

* In the Preface to "The Dockyards and Shipyards of the Kingdom" there was published the following proposal of Mr. Samuda to Sir Baldwin Walker, and its reproduction here stands in need of no apology:—

"I would propose to construct vessels in conjunction with the Government dockyards; that is, for private contractors to build the entire of the iron hulls, and attach all the armour-plates to them, and to deliver the hulls in the dockyards, there to receive the wood decks, magazines, and the entire of the wood fittings; all of which might be prepared by the Government at the same time that the iron hulls were preparing by the contractors. By this division of the work, which for many reasons I imagine would be advantageous to the service, added to the simplicity of construction, which, looking most carefully into this matter for a long period, has enabled me to suggest that I

been long identified with the construction of the steamships of the Peninsular and Oriental Company. Recently a portion of the extensive vacant space between Messrs. Samuda's original premises and Mr Stewart's engine-works has been appropriated by the firm, and the first launch took place the other day. Usually Messrs. Samuda have in hand as much as 16,000 tons of shipping, and recently there were the *Prince Albert* cupola ironclad, and the *Tamar* transport for the Admiralty, two vessels for the fleet of the Peninsular and Oriental Company, two for the Viceroy of Egypt, and four for foreign governments and steam navigation companies of importance. The *Prince Albert* is entirely built of iron, and the topsides for 11 feet from the gunwale downwards is to be covered with $4\frac{1}{2}$-inch armour, resting on 18 inches of teak. The entire topsides will consequently be protected with armour, and for five feet below the water. The armament will be in four cupolas on Captain Coles's plan, and the deck, which will be covered with iron, is adapted to firing at an object within 40 feet of the side. Finally, the stem is of great strength, and will be adapted to running down. Placing this ship in the hands of Messrs. Samuda was one of the few good things that the present Admiralty have done, because Mr. Samuda is an enlightened and zealous advocate of cupola ships. The ship of war building for the Viceroy of Egypt is of a very useful class, being only 350 tons burthen, and drawing no more than 4 feet of water, although carry-

---

think I could undertake to produce three vessels in two years (4,000 tons each) if I were required by the Government to hold myself completely at their disposal, and I imagine that three others might be found who would supply an equal number, and thus the Government might secure frigates at the rate of six a year, which I imagine, looking to what the French Government are doing, is the smallest progress consistent with safety."

4.—VIEW OF MESSRS. JAMES ASH AND CO.'S YARD, CUBITT TOWN.

4.—VIEW OF MESSRS. JAMES ASH AND CO.'S YARD, CUBITT TOWN.

ing two rifled guns in fixed shields at the extremities. This useful small craft is fitted with twin screws, and by these screws the guns will be trained and the gunboat steered. Recently a cupola corvette has been laid down in the Messrs. Samuda's yard for the Prussian Government, of somewhat similar dimensions to the cupola ship built for Denmark by the Messrs. Napier last year. The yard of the Messrs. Samuda is adapted for the construction of ships of the largest class, possessing as it does a fine launching frontage, and being as it is well stocked with machinery.

<small>Messrs. James Ash and Co.'s Yard, Cubitt Town.*</small> If Messrs. Samuda's yard ranks high for the construction of steamships of the first class and ships of war, and Mr. Lungley's yard is remarkable for the combination it presents, that of Messrs. James Ash and Co. is notable for the fact that Mr. Ash has been the successful designer of many of the best ships afloat. For eleven years Mr. Ash was in the employment of Messrs. C. J. Mare and Co., for

---

* The following are the names of some of the many vessels built from the designs of Mr. James Ash:—

OCEAN MAIL VESSELS.

The *Pera* ⎫
,, *Delta* ⎪
,, *Mooltan* ⎬ Peninsular and Oriental Steam Navigation Company.
,, *Poonah* ⎪
,, *Nepaul* ⎭
,, *Ly-ee-Moon* ...... Messrs. Dent, China.
,, *Genova* ⎱ Genoese Transatlantic Steam Navigation
,, *Torino* ⎰ Company.

CHANNEL PACKETS.

The *Normandy* ⎫
,, *John Penn* ⎪
,, *Prince Frederick William* ⎬ Dover Mail Steam Packet Company.
,, *Empress* ⎪
,, *Queen* ⎭
,, *Alliance* ⎱ South-Western Steam Packet Company.
,, *Havre* ⎰
,, *Lady of the Lake* ⎱ Southampton and Isle of Wight Improved
,, *Lord of the Isles* ⎰ Steam Packet Company.
,, *Fantaisie* ...... Yacht for the Grand-Duke Maximilian.
,, *Immaculate Conception* ...... Yacht for his Holiness the Pope.
,, *Dolphin* ...... General Steam Navigation Company.

six years in the employment of the Thames Iron Works, and it was during the last of these years that Mr. Mackrow, the present designer of the Thames Iron Works, acted as assistant to Mr. Ash. Than Mr. Ash the Thames has no more accomplished naval architect, and his yard is admirably adapted for the construction of ships of all classes. As a practical appreciation of Mr. Ash's architectural skill, Arthur Anderson, Esq., the chairman of the Peninsular and Oriental Company, has entrusted him with the building of a yacht,—a hint not to be lost sight of by yachtsmen, nor by the Admiralty when in want of fast-going craft for despatch purposes. Although only beginning business a year past Michaelmas, several ships have been launched by the firm: among the number, two vessels for the Peninsular and Oriental Company; a steamer for the South Western Railway Company, intended for the Channel Islands station, which has realised a speed of eighteen statute miles an hour; a steamer of more moderate speed for the Australian passenger trade; and two paddle-wheel steamers of 770 tons to realise a speed of nineteen statute miles an hour. The yard has a considerable river frontage, with a depth of 650 feet; and the plant, the whole of which has not yet been fitted, is of the newest and best designs and workmanship. When the firm began business the iron shipwrights and the bricklayers began work together,—a circumstance that places in a strong light the immense resources of this country for iron shipbuilding. To enable iron ships to be built anywhere a few furnaces are only wanted, and with these, and tools and iron, yachts, first-class steamers, or iron-clad ships of war may be begun. Is it not, therefore, a marvel that in this iron country a single voice should be henceforth raised in behalf of wooden ships?

4.—VIEW OF MESSRS. JAMES ASH AND CO.'S YARD, CUBITT TOWN.

Chapter XV.

THE MERSEY SHIPPING INTERESTS.*

The Birkenhead Ironworks. Messrs. Laird Brothers have an establishment of great compactness and completeness. With a frontage of only some 900 feet and a depth of only some 600 feet, they present building and repairing facilities such as raise the firm to a high position in the shipbuilding world. At the extreme right of the works the backward cupola ram-ship is building on No. 4 slip, and on the adjoining, or No. 3 slip, the forward cupola ram was built. The first of these slips is 230 feet long, and the second 240 feet long. To the left of these slips there is a tidal dock; next to that, No. 4 graving dock, 440 feet long, 85 feet wide, and 22 feet 3 inches of water on a 20 feet tide; next to that, No. 2 and No. 1 building slips, the one 250 feet, the other 400 feet; next to these, No. 3 graving dock, 400 feet long, 75 feet wide, and 26 feet of water on a 20 feet tide; next to that, No. 2 graving dock, 200 feet long, 45 feet wide, and 18 feet 3 inches of water on a 20 feet tide; and finally, next to that, No. 1 graving dock,

---

* This is not to be taken as a list of the Mersey shipbuilding firms, for these are very numerous.

300 feet long, 40 feet wide, and 18 feet 3 inches of water on a 20 feet tide. The extreme left of the works is a timber yard with a travelling crane over it. Behind, and more or less adjoining these docks and building slips, are grouped the erecting shops, the fitting shops, the engineer and millwright shops, the pattern and boiler shops, the rigging loft, mould loft, smiths' shops, angle iron, frame bending, punching, and other sheds.

The lesson of the Messrs. Laird's Works. The lesson of the Messrs. Laird's works is that, for shipbuilding and repairing on the greatest scale—the 6,621 tons ironclad *Agincourt* for the Royal Navy being in course of construction in No. 3 graving dock—very little more room is needed than is usually found unappropriated in front of the official residences in the dockyards. The nearer the work is to the workshops the more effective is the labour of the workmen, and were this principle only acted on in the dockyards large tracts of ground would at once become available for other purposes. Within the area of 900 by 600 feet the Messrs. Laird can simultaneously construct from 16,000 to 18,000 tons of shipping—a quantity equal to the whole present contract ironclad tonnage for the Royal Navy. Between the Messrs. Laird and their workmen the utmost cordiality prevails, the convenience and comfort of the workmen being constantly kept in view by the firm.

Messrs. Jones, Quiggin, and Co., Shipbuilding Yard, Liverpool. This yard is very compact, forming nearly a square, with an area of eight acres. The capabilities of this yard may be shown by the following statistics:—The river frontage contains nearly 500 feet. Frame and circular saw-rooms, 80 by

30 feet. Deck planing machine-room, 100 by 50 feet. No. 1. Joiners' shop, above and below, 180 by 25 feet, fitted with machinery, model-room in connection, and foreman's office and timekeeper's house. 2. Joiners' shop, 80 by 35 feet, fitted with machinery of the most modern and approved kind. The offices, general and private, occupying a convenient part of the yard and easily approached, the upper part of which are devoted to drawing-offices, &c. Centre of yard—furnaces, angle iron shops for smiths' work, with drilling shed attached, a number of punching and shearing machines, occupying together an area of 170 by 100 feet. Also, in addition are five sheds for various purposes connected with shipbuilding. No. 1. 100 by 25 feet, for making wooden spars, with boat-loft over. No. 2. 200 by 50 feet, for making iron spars, with mould-room over. No. 3. 200 by 50 feet, fitted up with single and double fires for shipping and general smith's work. No. 4. 200 by 50 feet, fitted with fires, with drilling, punching, and shearing machines for angles, plates, &c. No. 5. 200 feet by 50 feet, containing furnaces and machines requisite for bending frames and various other purposes. At upper end of sheds are a range of stores, 165 by 20 feet, containing paint, plumbing shop, and general requirements for the yard. The work the firm have on hand is as follows:—Two iron ships of 600 tons, two of 1,200 tons, three of steel 1,200 tons, one do. steel 1,350 tons, one steel paddle steamer of 1,500 tons, 350-horse power, one screw yacht 108 tons, one paddle 400 tons, 150-horse power, one screw 1,500 tons, 450-horse power; amounting in all to more than 12,000 tons; besides fitting other vessels in dock, which have been launched recently. The firm were the first who made steel spars for ships five or six years ago; since then they have

built many steel steamers and other small vessels for the Nile and other places. They were the first to receive orders for building merchant ships and steamers from 1,200 to 1,500 tons. They were also the first who adopted the use of steel for the rigging of the ships they are now building, the ships having standing rigging, &c., of steel, besides masts and spars of the same material. Mr. Jones, the senior partner of the firm, is the patentee of the angulated principle for war vessels. The machinery is driven by four engines of 20-horse power. Number of hands employed in the yard 1,200.

Messrs. Thomas Vernon and Son's Iron Shipbuilding Yard, Liverpool. In this building yard there are now in an advanced state of forwardness five large vessels, of which the following are the capacities:—one 1,200 tons, one 1,000 tons, one 900 tons, one 800 tons, and one 700 tons. In addition, Messrs. Vernon are about to lay down keels for five other large vessels, of an average of 1,000 tons each; and they have in the Liverpool Docks, just completed, the fine clipper ship, *Robert Lees*, of 1,200 tons register, and the barque *Mount Vernon*, of 560 tons register. During the past year they launched nine large vessels, of an aggregate of 9,000 tons. It is hardly necessary to state that these vessels, built and building, are all A 1 12 years' class at Lloyds', and 20 years' class with the Liverpool Underwriters' Association.

The firm of Messrs. Thomas Vernon and Sons is one of long standing—the late Mr. Thomas Vernon having been the constructor of thirty iron barges for the Shannon navigation thirty years ago. Up to 1843 Mr. Vernon had built and launched thirty-seven vessels, chiefly of iron, and some of them steam-vessels of considerable power and tonnage; and from 1844 to 1861,

THE MERSEY SHIPPING INTERESTS. 305

the firm built and launched no fewer than 109 vessels of an aggregate measured capacity of 46,000 tons. Besides this large amount of tonnage, the firm have had their hands more or less full with various engineering works; the great iron landing-stage at Prince's Pier, Liverpool, the iron landing-stage at Woodside Ferry, Birkenhead, and the iron low-water basin stage, Birkenhead, being the more prominent and important of these. Messrs. Vernon are also the constructors of the two first of Mr. Bourne's steam train of barges for the navigation of the Indus, as well as barges for the Ganges; and it will be remembered that they restored the Great Britain steamship after the unfortunate stranding in Dundrum Bay.

The dimensions of the yard are large, and the machinery and plant such as give it rank with the large private shipyards of the kingdom. From south to north the yard stretches along the river margin 486 feet, and from east to west 323 feet. In the central portion of the east side are situated a commodious range of writing and drawing offices, and contiguous the large drafting room in which the ribs and framing of the different vessels are drawn to the full size. The southern margin of the yard is occupied by the smiths' shop, which is 230 feet long by 40 feet wide. This shop, in which is turned out the smith's work of the yard, contains twenty blast hearths, and is amply furnished with all the tools required in the preparation of heavy ironwork, including the making of stern and rudder posts and the scarphing of keels. It also contains a steam-hammer of considerable dimensions and weight. Just beyond the smiths' shop, and still more south, is the shop and yard for bending the ribs and other portions of the framing of first-class iron ships. This shop is provided with the requisite extent of perforated iron floor. The premises contain, in addition to two sets of large rollers for bending or flattening rods or plate iron, the usual punching, drilling, and trimming machines of all the first-class yards.

Messrs. Vernon have also ironworks of a very large extent on the opposite side of the Mersey at Tranmere, and it is at these works where the landing-stages above referred to were constructed. The number of workmen employed by the Messrs. Vernon, at their two establishments, is about 1,300.

W. C. Miller's Shipbuilding Yard, Toxteth Dock, Liverpool. W. C. Miller's shipbuilding yard covers about 350 feet by 360, and possesses a river frontage of 350 feet. In this yard was built the *Oreta*, now called the *Florida*, and the *Alexandra*, about which so much stir was made relative to its seizure, which case was decided in favour of the defendants. The firm have been established ten years, and the experience of Mr. W. C. Miller, acquired by several years' occupation in the Devonport Dockyard, is a guarantee of knowledge and capability in building vessels, more especially for war purposes. In addition to the gunboats built by Mr. Miller during the Crimean war, the Government since that period have intrusted him with contracts for two despatch boats, and two large class gunboats. During the last twelve months he has built the first and only iron sailing vessel with an iron deck on Harland's patent principle, named the *Huddersfield*. This ship has also iron fore and main mast, top-masts, lower and double topsail yards, iron mizen-mast, top-mast, and top-gallant-masts, all in one. The vessel is built on an entirely new principle, combining all the advantages which can be derived from the substitution of iron for wood. Among the number of vessels built in the yard may be mentioned the *Phantom*, which was constructed entirely for running the blockade. She has made the passage from Liverpool to Madeira in six days nine hours. This vessel is also built entirely of steel, thus showing the advantages of vessels of that material. During the last eighteen months Mr. Miller has turned out twelve vessels of various sizes—steamers and sailing vessels. There are at present three vessels on the stocks, besides orders for several others. The three vessels on hand are, one for the Calcutta trade of 1,000 tons, nearly ready for launching. This ship is for Messrs. Prowse and Co., large shipowners of Liverpool, who have now given their attention to the substitution of iron vessels for their trade, as being more economical in every point of view. No. 2 vessel is a powerful iron dredge for the Liverpool Mersey Docks and Harbour Board. No. 3, a screw steamer for abroad. The machinery for the screw steamer and dredger are constructing by Messrs. Fawcett, Preston, and Co. This dredger is the second Mr. Miller has built for the Mersey Docks and Harbour Board. In the yard are various workshops for the purposes of shipbuilding. The extent of Mr. W. C. Miller's business, and which is daily increasing, will require larger and better-appointed premises. The advantages derived from the manufacture of iron on the spot are well attested. Mr. Miller has the Mersey Steel and Iron Works in close proximity to his yard.

The Britannia Engine Works, Birkenhead. These works are situated in the immediate vicinity of the great float, a short distance from the 60-ton crane and graving docks, and directly opposite the new chain test works, in a line with Duke-street. The railway passes through the works, being an extension from the London and North-Western and Great Western Companies' goods stations, branching off on the margin of the docks and float, and affording facilities for goods traffic to all parts of the kingdom. The works were commenced by Mr. James Taylor (under the firm of James Taylor and Co.) in the year 1852, in Cathcart-street, when several large iron structures were built for Australia; engine and boiler work being the principal feature in the establishment. The present workshops cover an area of about $2\frac{1}{2}$ acres, one-third of which is taken up by an iron and brass foundry, and the remaining portion the engineers' and boilermakers' departments, consisting of turning, erecting, and smiths' shops, and boiler yard, with offices. The works are well adapted for a large class of general engineering and boiler work, but hitherto have been chiefly devoted to the manufacture of steam machinery for hoisting purposes with double engines, for which a patent was taken out by Mr. Taylor, the proprietor of the works, in 1852, and since which they have been introduced into most of her Majesty's dockyards, where they are extensively used as labour-saving machines, into many private yards both at home and abroad, as well as yards of foreign Governments. They are also well adapted for carrying on the manufacture of marine engine and boiler work, having extensive and commodious sheds for the latter, fitted up with overhead travellers running the full extent of the shops, and other powerful appliances requisite for carrying on work of the heaviest class.

Woodside Graving Dock Company (Limited), Birkenhead. These premises are situated between the Monks and Woodside Ferries, covering a space of over 15,000 yards; the largest dock has an entrance of 85 feet, which is wider by 15 feet than any on the Liverpool side. There is 22 to 23 feet of water on the blocks on the highest spring tides; the other docks are smaller. Here repairs, both for wood and iron ships, are carried on, the machinery for each being on the premises. In the yard the iron masts for H.M. ships *Royal Oak* and *Caledonia* were made. The firm are now employed making masts for others of H.M. ships. In the large dock the late Pasha of Egypt's royal yacht *Fied Gehard*, paddle steamer, 82 feet wide, came in under her own steam, was blocked, and lengthened 70 feet without having to remove either engines or paddle-wheels; the length was 387 feet. The docks are built on and cut out of the red sandstone, and could be deepened to any extent for greater draught, if required, without disturbing the side walls, for a mere trifling outlay. The large dock is peculiar in construction, taking in three ships of 1,000 to 1,400 tons each; the firm have blocked five vessels in it on one tide; it is a wet or dry dock at pleasure, as they can retain the water at will. The other docks work with gates, as usual, simply excluding the water. There is little or no tidal current for 600 or 700 feet from the wall; in fact, what little there is always runs ebb, *i.e.* north, close to the wall. The docks are under the management of Mr. W. Ashley Clayton. These docks and those of their neighbours, the Messrs. Laird and Co., are the new private docks in the Mersey, the rest being under the control of the Mersey Docks and Harbour Board. Repairs, not building, are the primary considerations of the company, and for this the docks are advantageously laid out; all materials and workshops being within its own walls, saving both time and labour.

Since the above was written the company have taken the adjoining premises (7,500 yards) for iron ship building, having the London, North Western, and Great Western rails on the quays.

G. R. Clover and Co., Liverpool, and Private Graving Docks and Building Yard, Birkenhead. This firm have an extensive shipbuilding establishment in Birkenhead, in a direct line with Messrs. Clayton, Laird, and Co.'s. The repairing yard, with every convenience, is on the Liverpool side of the Mersey, at 1, Baffin-street, Queen's Graving Dock. Within the establishment are manufactured iron and steel masts, all kinds of ship ironwork, and every convenience for repairing ships in graving dock. The firm have been established since 1825, and have had considerable experience in the constant repairs of ships of all class of tonnage. The Egyptian Frigate, *Schea Gehead*, was built and lengthened by this firm.

The establishment at Birkenhead consists of three graving docks, which will accommodate about 5,000 tons of shipping at one time, and there is every facility on the premises for executing all kind of repairs in wood or iron. There is one iron ship nearly finished of 1,000 tons, and another of 1,200 tons nearly in frame. Since the establishment was opened in September, 1856, upwards of 490 vessels have been repaired, &c., in these graving docks, the tonnage of which amounts to 310,000 tons.

The length of blocks for the accommodation of vessels for repairs is as under:—

        No. 1.—Length of block 700 feet.
        No. 2.   ,,   ,,   280 feet.
        No. 3.   ,,   ,,   180 feet.

Vessels of any size can be docked in the above docks.

Sandon Graving Docks, for repairing, boat-building, &c., similar to the yard in Baffin-street, only on a minor scale. The Baffin-street yard has been established for fifty years.

<small>Cato, Miller, and Co. Brunswick Forge and Ironworks.</small> Messrs. Cato, Miller, and Co. supply many of the large iron shipbuilders, both in Liverpool and other parts, with forgings and iron masts. They have a factory capable of making and testing the largest anchors and cables.

Four large steam hammers, principally intended for manufacturing the forgings required for iron ships and engines, are used in making iron masts and yards.

The river frontage is 500 feet, and the yard is nearly six acres.

No. 1.—Large dock, 400 ft. long, with 80 ft. of entrance.
No. 2.—Dock, 300 ft. long, with 36 ft. of entrance.
No. 3.—Dock, 200 ft. long, with 35 ft. of entrance.

The draught of water in these docks at spring tide is 22 feet. There are extensive rooms for block-making, joiners' shops, mould lofts, fitting shops, smiths', machinery for executing iron ship-building, saw-pits and saw-mills, and every convenience for wood as well as iron ship-building.

Messrs. W. H. Potter and Co.'s Shipbuilding Yard, Baffin - street, Queen's Dock, and Blackstone-street, near Sandon Graving Docks, Liverpool. The Baffin-street yard has every accommodation for the building of iron vessels; and for the extensive business carried on, the capabilities of the yard for ship-building are most admirably adapted. The workshops, moulding-rooms, &c., are conveniently situated. The river frontage is of sufficient length for building three vessels at the same time, and the depth of water is adequate for the launching of any vessel of considerable tonnage, or in fact of any amount of tonnage. The length of the yard is upwards of 600 feet. No. 1, moulding-room, is 130 by 40 feet. No. 2, room for wood machinery, 86 by 26 feet. No. 3, engine-house, 30 feet in length. No. 4, angle-iron smithy, contains twenty fires. Ships' smithy; eighteen fires, with steam hammer. Forge for large work, ten fires. There are also several shops containing all kinds of machinery well adapted for iron ship-building. At present on the stocks are two large iron ships, the tonnage of which amounts to upwards of 2,500 tons. The yard is also adapted for the extensive repairing of wood and iron ships. The firm have also works at the north end, convenient to the